How to Live Comfortably for Several Years in a Hostile Wilderness Environment

Robert Wayne Atkins, P.E.
(Grandpappy)

Practical Advice
on How You Could Live
a Relatively Normal Life
in the Middle of a
Hostile Wilderness Environment.

For more information please visit: https://www.grandpappy.org

How to Live Comfortably for Several Years in a Hostile Wilderness Environment
Robert Wayne Atkins, P.E. (Grandpappy)

First Edition published by Grandpappy, Inc.

Practical Advice on How You Could Live
a Relatively Normal Life in the Middle of a
Hostile Wilderness Environment.

ISBN: 978-0-9850358-2-2

Printed in the United States of America.
10 9 8 7 6 5 4 3 2 1

Preface to

How to Live Comfortably for Several Years in a Hostile Wilderness Environment

There are two possible wilderness survival scenarios:

1. You become unexpectedly stranded in a hostile wilderness environment and you need to survive until you are rescued or you gradually make your way back to civilization.
2. You intentionally enter a hostile wilderness environment and you intend to live there for an extended period of time.

All the survival literature and the survival manuals that I have read focus exclusively on the first scenario above. This is also the primary focus of all military survival manuals. Therefore if the first scenario is the one you have in mind then you can easily find an overwhelming amount of information on that topic.

However, the second scenario is uniquely different. Instead of becoming unexpectedly stranded in a wilderness area, you intentionally enter the wilderness area with the intention of living there for several years. This strategy is **not** discussed in wilderness survival manuals because their primary focus is on how to survive until you can escape from the wilderness area and return to civilization.

The purpose of this book is to discuss the second scenario and to describe what you should take with you into the wilderness so you can successfully survive for an extended period of time. Some of these items are the same things you would take on a weekend camping trip but most of them are uniquely different. The reason they are different is because at the end of a weekend camping trip you intend to return to civilization instead of remaining in the wilderness for several years.

This book is **not** about "backpack survival" where a person survives in the wilderness using nothing more than the items he carries on his back. This book assumes you drive a fully loaded vehicle into the wilderness.

The reason I believe I am uniquely qualified to discuss this particular subject is because I spent approximately 6 months living in the backwoods of Maine with my wife and our three small preschool age children from June to November of 1975 while I was building a rustic log cabin using pine trees that grew on the 12 acres of land we had purchased. After returning to "civilization" in late November of 1975 I continued to study about pioneer life and wilderness survival.

During the past forty years I have personally tested a wide variety of recommended "textbook survival procedures" and I have discovered which ones are truly reliable and which ones are almost completely worthless from a practical perspective. Therefore I have accumulated a tremendous amount of practical hands-on knowledge on this topic. And it has all been tempered by my real-life experience of living in a hostile wilderness environment in the backwoods of Maine during the summer and fall of 1975.

Therefore the suggestions in this book were not simply copied from a survival manual. Although some of the information in this book can be found in some survival manuals, the vast majority of the information in this book is from my own personal files and it is based on my own personal life experiences.

One of the advantages of receiving wilderness survival instruction from an old man is that he no longer has the desire to prove that he can conquer the wilderness with a good hunting knife and the will to survive. An old man more fully appreciates the comforts of civilization and the advantages of modern technology. Therefore he will employ as much of that technology as is reasonable to make his family as comfortable as possible in the wilderness.

Respectfully,
Robert Wayne Atkins, P.E. (Grandpappy)
January 1, 2014

Table of Contents

Chapter **Page**

Chapter One

Introduction to Long-Term Wilderness Survival

The basic concept of long-term wilderness survival is a simple one. The real challenge is in its smooth efficient successful implementation. And that is where my experience will prove its true value in the long run.

The suggestions in this book will help you to strategically evaluate the things you probably already own and separate them into two categories:

1. things that would be truly useful in the wilderness, and
2. things that would not be useful in the wilderness.

For example, which of the following two items would be more useful for **long-term** wilderness survival?

1. A microwave oven.
2. A tent.

Both of the above items are approximately the same size and weight (when the tent is folded up for transportation). But each of the above items has a totally different function.

If you answered the microwave oven then you are correct. A tent would be far more useful for **short-term** wilderness survival but the microwave oven would be preferred for **long-term** wilderness survival. The reasons for this choice will be explained in detail later in this book.

I will also discuss a variety of traditional camping equipment items and explain:

1. which camping items are really practical for long-term wilderness survival, and
2. which camping items are not practical and should be avoided.

Finally I will discuss many items that are not traditional camping equipment but which are truly wonderful things to have and which could therefore make a really big difference in your comfort and survival in the wilderness. One of these items is a hand-operated meat grinder that will allow you to convert the occasional tough wild game meat into hamburger consistency to make it easier to eat and digest. This is just one example. There are many more.

The reason this book could be critical to your long-term wilderness survival is because of three related factors: (1) weight, (2) size, and (3) function. Some items are small but they weigh a lot (ammunition). Other items are large but they do not weigh very much (sleeping bag). Regardless of how many trips you make back and forth to stock your wilderness retreat you will have maximum storage space constraints in addition to weight constraints. A careful strategic selection of each item based on its: (1) weight, (2) size, and (3) function could easily make the difference between whether you live or die. Your choice of items will also determine whether or not you will be able to thrive in the wilderness instead of just barely being able to survive.

This book does not duplicate the information in my other books. This book discusses information that is not in my other books and that is specifically related to long-term wilderness survival. However, a small portion of the information in my other books is sometimes repeated in this book in order to provide a brief overview about a specific topic. After presenting the brief overview about a topic this book will then suggest that you read one of my other books for a more detailed discussion of that topic. This was necessary in order to keep the size and the cost of this book reasonable. However, you do not need to actually purchase my other books because a lot of the information in those books is also freely available on my web site. In Appendix Three I have included a list of direct links to the free information on my web site so you can access that information immediately if you wish. However, after you see the quantity of free information that is available on my web site then you may decide it is actually cheaper to purchase one of my other books instead of printing hundreds and hundreds of pages of free information yourself.

The Following Major Topics Will Be Discussed In This Book

1. How to form a proper perspective about long-term wilderness survival.
2. How to select a suitable wilderness environment.
3. A discussion about your current possessions so you will know which ones you should take with you into the wilderness.
4. A complete list of additional items that you should purchase and take with you.
5. A list of firearms and knives that are absolutely necessary for wilderness survival.
6. How to replenish your water for drinking and personal hygiene purposes.
7. Food recommendations and how to replenish your food supplies from nature.
8. Practical cooking options in a wilderness environment.
9. Propane as a cooking and heating fuel.
10. Electrical power options for a wilderness shelter.
11. Shelter alternatives for long-term wilderness survival.
12. Equipment waist belts and waist packs.
13. Some additional issues about long-term wilderness survival.
14. Some practical strategies for long-term wilderness survival.

There are also four Appendixes at the end of this book as follows:

Appendix One: Basic Construction Plans for an Underground Home.
Appendix Two: Practical Wilderness Survival Information Available on the Internet.
Appendix Three: Links to Free Survival Information on Grandpappy's Web Site.
Appendix Four: Recommended Books for Basic Wilderness Survival.

Chapter Two

A Different Perspective About Wilderness Survival

This will not be an exciting or a fun adventure. This will not be a leisurely camping trip. It will be 100% hard work. Your body will age about two-years for every one-year you spend in the wilderness. After the first week in the wilderness all the romantic notions you had about wilderness living will be completely forgotten. You will feel loneliness like you have never experienced before. You will experience new fears and new doubts. You will seriously consider giving up and returning to civilization regardless of what might be waiting for you there. If you are emotionally prepared for these negative feelings and for all the hard work you will have to do then you might actually survive for several years in the wilderness. However, if you are not prepared then there is a very good chance you will simply give up before one-month has passed.

Before you enter the wilderness you should consider the following:

Identification: If possible renew your driver's license for the maximum number of years allowed by your state so your driver's license and birth certificate will be legal proof of your identity for the next 5 or 10 years. If you have a passport then take it with you also.

Planning: Do not plan for an extended camping trip. Do not think about all the neat camping gear and tents you could buy in order to have a really great camping adventure. This is the wrong perspective and if you follow this line of thinking then you will probably not survive for several years in the wilderness.

Instead you should plan to move into the wilderness. In other words, think about this from the perspective of simply moving to a new home in a new area but with a few differences. There will be no electricity, no running water, no sewer, and no central air-conditioning or heat. There will be no hot fresh pizza deliveries.

However, to the extent possible, you would still like for your life in the wilderness to closely resemble the life you are now living. The major exceptions will be that you will be working for yourself and you will not be going to a "regular job" every day. You will have essentially no bills to pay, except annual property taxes. You will not be buying things at the store on a weekly or monthly basis. You will be providing for all your own needs. You will be providing for your own security and self-protection.

But you will not be doing these things the way your ancestors did 200 years ago. You will not arrive in the wilderness inside a covered wagon that is pulled by a team of mules. You will not be sleeping on the ground under the wagon at night. You will not be shooting a single-shot black-powder rifle or musket.

Vehicle: Instead you will arrive in a modern vehicle; preferably a mini-van or a standard size van. You will be bringing many modern conveniences with you including modern firearms. Your supplies will be neatly stacked on the floor of your van and you will have a nice mattress, either full or queen size, on top of those supplies. (Note: Vans and mini-vans both have very low floors and a lot of space between the floor and the roof of the van.)

A Typical Mini-Van

Bedtime: A good mattress is one of the most important items you can take with you into the wilderness. All you need is the top mattress. You don't need the lower foundation. If you can get some good sleep every night then your body and your mind will be refreshed and the following day you will be able to make better decisions and do more work. However, if you don't have a good

A Queen-Size Mattress

mattress then you will not get a good night's rest and you will be miserable during the night and during the day. Any size mattress is better than no mattress. If you have the space then a full-size or a queen-size mattress would be ideal. But if space is limited then a twin-size mattress would also be fine for one person. A full-size mattress will fit flat inside the back of a mini-van above the wheel wells, and it can be supported on a piece of plywood on top of your storage buckets or plastic totes, and you will still have enough room to sleep comfortably on top of the mattress.

Your Temporary Home: While you are constructing your new home you could sleep inside your van on your mattress, on clean sheets, under a warm blanket or under a cold-weather sleeping bag. You will be able get a good night's sleep every night. You will be out of the wind and rain. If you do not spill any food or beverages inside your van then you will not be bothered by bears or by snakes. And you will not have ants or other insects crawling all over you during the night. (Note: One problem with a van as a permanent home is that you can't stand straight up inside the van.)

Your New Future Wilderness Home: You should not construct a traditional home or a log cabin. Instead you should build an underground home into the side of a steep hill or a "wilderness cabin cave." These options are discussed in the shelter chapter of this book. In you have either one of these types of homes then you will not have any significant heating or air-conditioning issues because your new home will have an inside temperature that varies between 50°F to 60°F (or 10°C to 16°C) all year. This means you will not need to cut a huge stack of firewood to comfortably survive the winter. Since you will not need a lot of firewood then you will have reduced your total annual workload by at least one-half. On the other hand, if you live in a traditional home or a log cabin then you will need to cut, split, haul, and stack enough firewood to survive the winter and this activity will compose at least 50% of your total annual workload. You can eliminate this unnecessary heavy workload by constructing the most energy efficient type of home.

Utilities: You will have a solar shower bag that you can fill with water each morning and then you can place it in the sun so the sun will heat your water for free. You will have a simple solar power generator consisting of two or more small solar panels, two or more deep-cycle 12-volt batteries, and an inverter. You will be able to use the sun to charge your deep-cycle 12-volt batteries for free. You can then use the power inside your 12-volt batteries to charge your rechargeable flashlight batteries, your cell phone, and your portable computer, and you will also be able to operate a few other small appliances for short periods of time.

Summary: Your life will have changed. On the days you work you will be physically exhausted at the end of the day but you will be able to take a warm shower. And you will be living inside a manmade cave or a home built into the side of a hill. But you will still have music, DVD movies to watch, a radio to listen to, books to read, and a really nice mattress to sleep on every night. And you will have food to eat that you provided one way or another.

When you read the above it all sounds so easy. It is not. It is not easy. It is not easy. It is very, very hard work. And most of the people who try it will not succeed. The wilderness will gradually wear them down and they will give up. If you know this before you begin then you will have a much better chance of being one of the very few people who can actually live and thrive in a wilderness environment.

Chapter Three

How to Select a Suitable Wilderness Environment

Temperature: The first major issue that should be considered is temperature. There is a very good reason why there are a lot of untamed wilderness areas in the northern United States. Those areas are not suitable for long-term wilderness survival. The summer growing season is too short and the winters are too long and too cold. Remember that your objective is not to simply get to a wilderness area. Your primary objective is to survive and thrive after you get there.

On the other hand, any wilderness area in the middle or southern United States has the potential for long-term wilderness survival. The summer growing season is several months in duration and the winters are cold but survivable because of their shorter duration and their milder temperatures. The important issue is not the lowest possible winter temperature but the average winter temperature for each month from November through March.

Rainfall: The second critical issue is annual rainfall. Look at the rainfall data for any area you are considering and determine what the average rainfall has been during for each month for the past few years. If the area has a history of long and sustained droughts then eliminate that area from further consideration. If the area has a history of occasional flooding then eliminate that area from further consideration.

Safety: The third critical issue is the area's recent history of earthquakes, hurricanes, tornados, volcanic eruptions, and forest fires. If the area has a bad safety record then eliminate it from further consideration.

Altitude: The fourth critical issue is altitude. The wilderness area should be at least 300 feet above sea level but not more than 5,000 feet above sea level. Areas at low elevations could easily be flooded during a catastrophic hard times event, and elevations that are too high are colder and the winter weather arrives sooner and lingers longer.

Wild Game: The fifth important factor is a reasonable wildlife population of edible wild game animals such as rabbits, opossums, groundhogs, raccoons, muskrats, and beavers. Other desirable wildlife would include deer, black bear, and game birds such as turkeys, pheasants, and ducks.

Predatory Animals: The sixth factor is predatory animals. An area with fewer meat-eating predatory animals is better than an area with a lot of meat-eating predatory animals. Meat-eating wild animals will be in competition with you for the available wild game and they may also consider you to be an acceptable meal. Therefore I recommend that you carefully research the currently available information about the predatory wild animal population in any wilderness area you are considering. For example, almost all wilderness areas have bears. But there is a big difference between a black bear and a grizzly bear. A black bear will usually not attack a person unless it is provoked, or it believes you are a threat to its cubs, or it is starving to death and you are the only meal that is within its grasp. On the other hand, a grizzly bear will not hesitate to immediately attack a person and then eat that person. In some areas the wolf has once again become a dominate predatory animal and a pack of wolves will not hesitate to attack people. Other predatory animals include the cat family, such as cougars, panthers, mountain lions, wildcats, and bobcats. Wild boars can also be dangerous. Therefore you should carefully consider the types of dangerous wild animals in the wilderness area you are considering along with the estimated number of those animals in that area, which is usually expressed as animals per square mile of wilderness.

Wolves

Poisonous Reptiles, Spiders, and Insects: The seventh factor is dangerous reptiles, spiders, and insects. All wilderness areas will contain some of these poisonous creatures. The primary areas of concern are their types and their numbers, which is usually expressed as the number per square mile of wilderness. You should be able to accurately identify all of these poisonous creatures on sight so you can avoid them. You should also know the proper first aid treatment if you are bitten or stung by one of these creatures. This is not the type of information that you want to waste time trying to find after you have been bitten or stung. This is the type of information you should have memorized before the accident happens so you can immediately execute the proper first aid treatment. You should also either immediately call for help on your cell phone or you should be immediately taken to an emergency medical facility for the correct treatment that is performed by a licensed medical practitioner.

Population Density: The eighth important factor is population density. A wilderness area by definition will have a relatively low human population. However, if the wilderness area is within 50 miles of a major city then the wilderness area could easily and quickly be overrun by the people who live in the big city if there is a future catastrophic hard times event. Therefore wilderness areas that are further away from major metropolitan areas are preferred to areas that are near large cities. However, it should be noted that almost any area east of the Mississippi River will have a higher population density than any area west of the Mississippi River. But this does not mean that you should automatically eliminate any wilderness area east of the Mississippi River. The important issue is to think in relative terms. Any area east of the Mississippi River should be compared to other similar areas east of the Mississippi River and any area to the west should be compared to other areas in the west.

Non-Critical Issues: Some of the things that are not critical in the selection of a suitable wilderness area are the following:

1. The distance from your current location to the wilderness area.
2. The proximity of a national forest or a wildlife protection area.
3. The distance to the nearest small town.

After you have selected a wilderness area that meets the above critical criteria then you must make a decision on a specific parcel of land. The size of the parcel will be determined by your budget. Unfortunately specific investment advice, real estate advice, and legal advice goes far beyond the original purpose of this chapter.

Chapter Four

What Should You Take With You into the Wilderness?

Current Possessions: Many of your current household possessions will probably be of very little or no value in the wilderness. This includes your big-screen televisions, your big desk top computer and monitor, and most of your electrical kitchen appliances. Therefore if you are serious about moving into the wilderness then you should consider selling everything you will not need. Or you should give it to a relative who would truly appreciate it, or you could donate it to a "Thrift Store." If some of your things are old, worn out, and relatively worthless then you should take them to the trash dump. On the other hand, you could rent a storage facility and store your furniture and possessions. The small rental units sometimes change ownership and managers so a standard warehouse that gives warehouse receipts may be a better choice.

For this to be a practical alternative the "true current market value" of your possessions must significantly exceed their actual long-term storage costs. An item's true market value is what you could sell it for today and not what you originally paid for the item. You may have paid $900 for a new sofa ten years ago but if you tried to sell it today you probably would be offered $100 or less for it. As an example, assume that you could sell almost everything you have for about $1,200 cash and assume the cost to store all of those items in a warehouse would be $100 per month. Based on these numbers if you rented a warehouse for more than 12 months then the warehouse rent would exceed the true market value of your possessions. In this situation you would probably be better off selling your possessions and investing the proceeds in items you will desperately need to survive in the wilderness. However, if the true current market value of your possessions is about $25,000 then you will need to think very carefully about whether or not you should store those items or convert them into cash.

If you do not have a lot of things then you may be able to find a close relative who will store your items for you in his or her garage or storage building or barn. Generally large bulky items take up more space and cost more to store. Smaller items are frequently more valuable and take up less space and therefore cost less to store. Therefore you may want to store the small stuff and sell the big stuff. This is a very serious issue and you will need to make a decision about your current possessions that is appropriate for your particular situation.

Clothing: Buy or select dark color clothes or camouflage patterns to blend into the wilderness. White and light color clothes can be easily seen from a long distance away and you will be quickly noticed by everyone and by every wild animal. Your objective is to blend in and become almost invisible. Therefore select very dark colors for your outerwear and your underclothes. For example, black, dark blue, dark brown, dark green, or dark charcoal gray would be good choices for everything from socks and underwear to waterproof jackets. The only problem with having 100% camouflage clothes is that they quickly become boring to look at and they attract too much attention when you are in a town. When you occasionally visit a town it is better to have simple dark solid color clothes so you can blend in with everyone else.

You should also dress in layers so you can add layers or remove layers depending on the changing temperature. If you only have thick heavy-weight clothes then will not be able to do this. For example, in the winter you should put on a dark color t-shirt and dark underwear. Then add dark thermal long underwear uppers and lowers (you can dye white cotton a darker color). Then add dark socks. Then add dark pants and a dark long-sleeve shirt. Put on some dark shoes or boots. Then add a dark waterproof jacket. Then add a dark ski mask and then cover that with a dark hat. Then add soft warm dark gloves. You would now be ready for winter weather. If it got really warm on a winter day then you could easily remove an item or two or three and remain comfortable and not sweat.

Perspective: Resist the temptation to become "true pioneers." In other words, take advantage of every possible labor saving device and tool that you possibly can, such as butane lighters, fish hooks, sewing needles, steel knives, modern firearms, water filters, cold weather sleeping bags, etc. Long-term wilderness survival will be challenging enough without making it more complicated than it needs to be. This will not be a simple weekend excursion or a training exercise. You will not be graded. You will not be trying to impress anyone. You will simply be trying to survive in as comfortable a fashion as possible without exposing yourself to unnecessary risks, or to physically exhausting or time-consuming work, if it can be avoided. For example, you could invest several hours trying to start a fire using a primitive fire starting method. Or you could start a fire in a few seconds with a match. If your life depended on the outcome then which method would you prefer?

Sleeping: As you select your items you should think in terms of a long-term home as opposed to a short-term camping trip. On a camping trip a nice air-mattress or cot is fine for a few days. But it is totally unacceptable for several weeks, months, or years. Therefore, if possible, you should have a regular mattress to sleep on at night. You should also have some good sheets and a good pillow. You could cover your bed with a thick sleeping bag to conserve your natural body heat. If you can get a good night's rest then you will be better prepared to meet the challenges of each new day. But if you do not get a good night's rest then you will be exhausted the following day.

Water: Water will be one of your most important considerations and you should have thought about this issue in great detail ahead of time. My recommendation is that you have two five-gallon water transport containers so you can take some water with you into the wilderness. You should also have several good tarps so you can replenish that water by capturing rainwater. When it rains you can take the day off and relax, and gradually refill all your water storage containers (heavy-duty plastic totes) with rainwater.

Chapter Five

A Complete List of the Additional Items You Will Need

The following list of items would be absolutely essential to your long-term wilderness survival:

High-Quality Heavy-Duty Plastic Totes (or 6-gallon food grade buckets): These totes should be waterproof, insect proof, and have tight fitting lids. Pack your food and survival gear inside these totes so you can easily and conveniently transport your possessions. As you gradually consume your existing food supplies these totes can be used to store your future food supplies such as meat jerky, dried fruits and vegetables, pemmican, acorns, and nuts. These plastic totes can also be used to store rainwater each time it rains. Resist the temptation to buy the cheaper thin-wall plastic totes with loose-fitting light-weight lids. These cheap totes will not survive normal wilderness living. They will crack, split, and tear due to normal handling and to temperature fluctuations.

Heavy-Duty Two-Wheel Dolly: You may need to transport your possessions from the road to your new wilderness home. A two-wheel luggage carrier or dolly with the largest possible wheels would be ideal for this task. Some hunting supply stores sell a special "wild game carrier" that has really big wheels for transporting a heavy game animal out of the woods. This type of carrier would be very useful for moving your possessions from the road to your new wilderness home. It would also obviously be useful for moving big game animals from deep in the woods back to your wilderness home. (You will also need a manual air pump.)

Furniture:

Your current full-size or queen-size mattress (but not the lower foundation).
(If you have children then a twin-size set of bunk beds and a ladder).
Two or more very comfortable pillows.
Two or more sets of good sheets (new or almost new) that fit the above mattress.
Blankets or quilts.
A cold-weather sleeping bag (rectangular shape and not a mummy bag).
Your current kitchen table and chairs (or a folding table and folding chairs).
Your current bedroom dresser, and chest-of-drawers, and small bedside end tables.
Your current sofa or a comfortable queen-size sleeper sofa (for children).
One or two small table lamps with 60-watt incandescent light bulbs and twenty extra 60-watt bulbs.
Curtains for a bedroom window and for a living room window.
A few framed pictures to hang on the walls (peaceful happy scenes).

Weapons:

Semi-automatic center-fire hunting rifle or assault rifle with a scope and a sling.
3,000 rounds of rifle ammunition.
Semi-automatic pistol (40 or 45 caliber) with two magazines.
Pistol holster with a compartment for the extra magazine.
1,000 rounds of pistol ammunition.
22LR Ruger semi-auto rifle with see-through scope mounts, scope, and a sling.
10,000 rounds of 22LR ammunition.
Beeman Air Rifle 22 caliber and 177 caliber Combo or equivalent (silent harvesting of very small game).
5,000 hollow point or pointed air rifle pellets (22 caliber).
Gun cleaning kits, patches, solvent, and lubricant.

Two fixed-blade stainless steel hunting knives in sheaths.
Two folding stainless steel pocket knifes.
Two Arkansas sharpening stones or 2 diamond sharpening stones or 2 Pocket-Pal knife sharpeners.

Tools:
Portable quality tool kit in a plastic carrying case (hammer, screwdrivers, wrenches, etc.).
Reasonable assortment of extra spare nails, screws, bolts and nuts.
25-foot retractable tape measure.
One "Leatherman Wave" multi-tool with belt sheath (carry when trapping or fishing).
Hand-operated manual drill and drill bits.
Crosscut saw that may be used by one person using one hand.
Heavy-duty metal file.
Hatchet with a stainless steel handle and a comfortable non-slip grip and a belt sheath.
Shovel, Hoe, Rake, Pick, and Axe -- all with strong fiberglass handles.
Heavy-duty pruning shears with 24-inch long fiberglass or metal handles.
Either a hide skinning tool or a tree bark removal tool (called a "draw knife").
Handheld garden trowel and handheld pruning shears.
Sledgehammer with one splitting edge for splitting firewood (future smokehouse).
Steel wedge for splitting firewood (future smokehouse).

Note: A convenient place to store many of your bigger tools is on a tool rack mounted behind the front door to your shelter. Your tools will be inside your shelter and beside your entrance door so you can easily get to them when you need them. And you can easily hang them up when you are done with them. You can keep them from rusting if you will wipe and dry them off before putting them on the rack after each use. You should also apply a thin coat of oil onto the metal parts at the beginning of each winter.

Clothing: (Note: All dark colors or camouflage patterns to blend into the wilderness.)
An extra leather belt.
Twelve dark color t-shirts.
Twelve dark color underwear.
Six pair of thermal long underwear uppers and lowers (note: you can dye white cotton underwear a dark color but you cannot dye synthetics).
Twenty-four pair of dark color thick wool-blend socks (or cotton-blend socks).
Twenty-four pair of nylon ankle high footies (wear under your socks to prevent blisters).
Six dark color long sleeve shirts.
Six brown or black denim jeans (work jeans and not designer label jeans).
A winter waterproof jacket with an attached hood (a removable zipper lining is preferred if available).
Two pair of quality "comfortable" waterproof hiking boots or work boots with reinforced toes.
Good waterproof hat with a wide brim that goes all the way around your head.
Rain gear: plastic jacket and plastic pants.
Eight pair of flexible durable leather gloves for manual labor.
Two pair of soft cotton gloves for warmth indoors during the winter.

One ski mask and one hat with ear flaps that covers your head and ears for warmth indoors and outdoors during the winter.
Mosquito head net for warm weather while working outdoors.
Hand sewing needle assortment for repairing your normal clothing (two packages).
Heavy-duty craft sewing needles for making buckskin clothing and moccasins.
Thread for repairing your current clothing (10 big spools of assorted colors).
Dark flannel fabric for constructing future clothing (at least 20 square yards).

Brown or black denim fabric for future clothing needs (at least 20 square yards).
Button assortment (2 small packages that contain 50 or more buttons per package).
(Note: A button-up fly was used before the invention of the zipper fly.)
Stainless steel high quality sewing scissors (at least 5-inch long straight sharp shear blades).

The cap in the picture on the right is the perfect hat for cold weather indoors and outdoors because it is waterproof. The ear flaps can be tied above the hat out of the way when they are not needed. But as it gradually gets colder you can lower the ear flaps and the back center piece of the cap, and you will be able to keep your ears and the back of your head warm. When you select a hat like this it should not feel tight on your head. You should be able to shift it a little from side to side without it pulling the skin on your face. However, it should not be so loose that it falls off your head when you bend over.

The cap in the picture on the right is a wool cap and it is also an excellent choice. However it is not waterproof so it would not be practical during wet weather. On the other hand, it is the perfect type of cap to wear when you go to bed at night if you sew some chin tie straps to the right and left sides of the cap. The hat will cover the front of your forehead, and your ears, and the entire back part of your head. In the old days everyone wore one of these "night caps" at night during cold weather. If you also wear some wool socks, some soft flexible warm gloves, and some thermal long underwear then you will be able to stay warm under your blankets even if the temperature inside your shelter gets really, really cold.

The gloves in the picture on the right are driving gloves. They allow you to easily flex or bend your fingers. They have grip strips on the inside fingers to help you gain control of the things you wish to pick up. And they are very comfortable. These are the gloves I wear indoors on really, really cold winter days and when I go to bed on extremely cold winter nights. They really do keep your hands warm. However, I do not wear these gloves for any type of work activity. Leather work gloves should be worn for normal work activities.

Personal Hygiene:
2 Portable solar-heated shower bags (four or five-gallon plastic bag with a shower hose) (can also be used to provide warm dish water).
120 Bars of white Ivory soap with no additives (use for bathing, laundry, and washing dishes).
6 Bars of Lava brand hand soap (use on those days when your hands get real dirty).
3 Bath towels, 3 hand towels, and 3 wash cloths.
6 Dish towels and 6 dish cloths.
120 Double rolls of toilet tissue (provide 100% of your needs -- no gas station or restaurant restrooms).
12 eight-ounce tubes of toothpaste (use a tiny amount of toothpaste each time you brush your teeth).
3 good quality toothbrushes (replace your toothbrush once a year or when it just wears out).
1,000 yards of dental floss.
12 disposable safety razors (shave before you make a trip into town).
Fingernail & toenail clippers, comb, hair brush, nail file, 2 good hand mirrors, tweezers, barber scissors.
Fingernail brush.

Folding camp toilet for outdoor use.
Bucket toilet seat for indoor use.
36-inch to 48-inch diameter round metal pan for indoor shower use (or a big plastic tote).

Toilet Tissue: Do not underestimate the value of toilet tissue to your personal health, hygiene, and peace of mind. It is one of the simple comforts of civilization that you will dearly miss if you don't take enough of it with you. You will need to provide for 100% of your normal toilet tissue requirements because you will not occasionally be using the toilet tissue at your place of employment, or at a gas station, or at a restaurant.

Nails: Always keep your fingernails and toenails trimmed in the wilderness. You will be using your hands a lot and long nails get caught on stuff and break or tear (very painful). Long toenails will wear holes through your socks. Sewing up the holes will extend the life of the socks for a short time but at the expense of foot blisters at each place you sewed up a hole. To avoid foot blisters, wear one pair of ankle high hose next to your skin and your regular socks on top of that.

First Aid and Safety:

First aid kit and alcohol wipes and anti-itch cream and insect repellant and sunscreen.
500 Bandages (mostly 3/4 inch or one-inch wide).
Oral antibiotics.
Over-the-counter medicines (Ibuprofen, diarrhea tablets, anti-itch tablets, etc.).
Two good pairs of quality safety glasses (clear lenses) (in hardware and sporting good sections of store).

Electrical:

L.E.D. high-power handheld flashlight.
L.E.D. solar power hand-crank portable camp lantern.
Solar battery charger and at least 16 spare rechargeable batteries (AAA or AA as appropriate -- read chapter eleven).
Battery operated world-band radio (Grundig G6 or Grundig S350DL).
Battery operated AM-FM radio (backup radio if the above radio malfunctions).
Between four to six 30-watt solar panels (or 60-watt solar panels).
Two, three, or four 12-volt heavy-duty batteries (marine or golf cart).
1200-watt or 2000-watt inverter (converts 12-volt DC to 120-volt AC).
25-feet medium-duty 14-gauge 120-volt three-prong electrical extension cord.
200-feet 14-gauge three-wire electrical wiring (for 120-volt and 12-volt applications) .
12-volt cigarette lighter 3-plug adapter for recharging battery operated items.
Your primary portable computer with a DVD Drive and a 12-volt adapter.
A backup portable computer with a DVD-Drive and a 12-volt adapter.
Your collection of your favorite DVD movies (no original plastic display cases -- just the DVDs inside a carrying case that holds one-hundred or more DVDs).
Collection of your favorite music stored on your computer and on a backup DVD.
Battery operated wall clock with hands that only shows the time of day.
Battery operated portable digital alarm clock with date, day of the week, and maybe the temperature.
Your current (or new) small microwave oven (1000 watts or smaller).
Optional: A cell phone or a satellite phone.

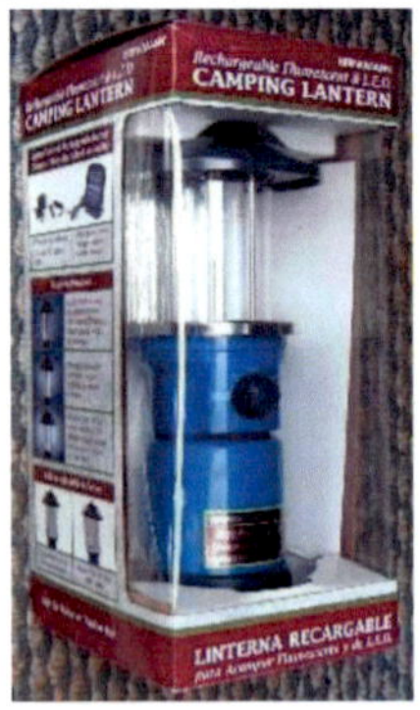

Miscellaneous:

6,000 matches, 12 butane lighters, and 2 magnesium fire starters.
2,000 yards of nylon or polyethylene twine (catfish twine with a minimum 150# breaking strength).
1,000 yards of 20 gauge metal wire.
100 yards of 14 gauge metal wire.
100 yards of half-inch thick rope.
Block and tackle (may be purchased on the internet).
Two good compasses.
Outdoor thermometer (mercury filled –60°F to +120°F or -50°C to +50°C) with barometer and humidity dials.
Indoor thermometer (mercury filled -20°F to +120°F or –30°C to +50°C).
Oven thermometer (+50°F to +450°F or +10°C to +230°C) (for future smoke house).
Refrigerator/Freezer thermometer (-40°F to +80°F or -40°C to +25°C) (for future root cellar).
Duck tape and clear plastic sheeting (at least 3 mil thick and 20 feet by 25 feet per roll).
Broom, Dust Pan, Mop.
2 Fly swatters.
4 Mouse traps.
Professional strength bug poison (from hardware store or pest control store).
2-Ton Epoxy Cement (two tubes that you mix together as you need it).
50-feet of vinyl coated clothesline wire and 100 clothespins and a clothespin storage bag.
Three heavy-duty dish pans just for washing dishes and one dish drainer.
Two heavy-duty dish pans just for washing your hands.
Four 50-foot garden water hoses (total of 200 feet of garden hose).
Two garden hose nozzles or two regular simple faucets that can be attached to one end of a garden hose.
Four gallons of 30W motor oil (synthetic preferred) -- for all your oil requirements.
Two gallons of Automatic Transmission Fluid (ATF) -- to clean your firearms.

Bicycle for future emergency travel purposes.
Family photographs carefully stored and protected (hard copies and digital copies).
Family heirlooms and printed genealogy information to pass down to your children and future descendents.
Legal Documents (Passport, Birth Certificate, Immunization Records, School Records, Deeds, Titles, etc.).

Books:

Nelson's NKJV Study Bible, 2005. (Note: Thumb-indexed bibles are very nice.)
SAS Survival Guide, Collins Gem, Wiseman.
Survival Gunsmithing, J.B. Wood, 1986.
Merck Manual of Medical Information, 2nd Home Edition, 2004.
The American Red Cross First Aid and Safety Handbook, Handal, M.D., 1992. (10 pages on childbirth.)
New Illustrated Guide to Gardening, Reader's Digest, 2000.
Big Book of Preserving the Harvest, Costenbader, 2002.
Joy of Cooking, Rombauer and Becker, 915 pages, 1975 Edition or before.
Back to Basics, Reader's Digest, 1997.
New Complete Hoyle Revised, Hardbound, 1991.
The Complete Book of Solitare and Patience Games, 225 Games, Morehead, 1983.
100 or more used assorted paperback novels (fiction, mystery, adventure, westerns, etc.).
Any other books you believe are important (business, technical, science, etc.).

Entertainment:

Four decks of plastic coated playing cards.

Writing paper, pens, pencils, and a manual pencil sharpener.

Five ordinary six-sided dice.

A reasonable variety of conventional board games, such as Monopoly, Life, Clue, Scrabble, Chess, Yahtzee, Chutes and Ladders, Candy Land, and any other games you currently own.

My book **"Ancient Board Games and Solitaire Games From Around the World."** Or you could download this game collection for free from my web site and print hard copies of all the games.

You should not be lacking for ways to enjoyably pass the time on rainy days or during the cold winter months.

Online Books:

Download and print a complete copy of the **Michigan State Trapper's Guide** and keep it in a 3-ring binder.

http://www.michigan.gov/documents/dnr/MI_Trapper_Education_Manual_82307_206561_7.pdf

Download and print a complete copy of the **Ten Bushcraft Books** and keep them together in a 3-ring binder.

http://tions.net/CA256EA900408BD5/vwWWW/outdoor~03~000

Chapter Six

Firearms and Knives that are Absolutely Necessary for Wilderness Survival

Your short-term and your long-term survival in the wilderness will be directly related to the type of firearms and knives that you take with you into the wilderness. This is a critical decision and your choice of firearms and knives for wilderness survival will probably not be the same choice you would make for personal self-defense in a civilized area. There is a huge difference between the wilderness and civilization and your decisions should be based on where you will be spending the vast majority of your time. This chapter will discuss firearms and knives from the perspective of long-term wilderness survival instead of a self-defense situation inside a populated area.

Firearms

The first issue that needs to be addressed is that this will not be a conventional hunting expedition where you and your friends enter the wilderness, shoot some wild game animals, and then return to civilization. On a conventional hunting trip you would probably be hunting a specific type of wild game animal and you could optimize your chances of being successful by selecting the most appropriate firearm and caliber of ammunition for that specific type of animal. If you decided to hunt a different type of wild game animal then you could change your firearm and ammunition based on the specific type of animal you would be hunting. In other words, you may have a variety of different firearms and ammunition that you could select from if you were going on a conventional hunting expedition

However, when it comes to wilderness survival you will not always know what type of animal you might encounter on a hunting field trip. And you probably will not know what type of wild animal may wander into your camp and decide to eat you. Therefore you will not have the luxury of being able to select from a huge collection of firearms. Instead you will need one firearm that can successfully do the job of killing any wild game animal that you might encounter. This will result not only in your survival but it may also result in some fresh meat for your cook pot.

In the wilderness you really need semi-automatic firearms. You do not need a bolt-action rifle, or a lever-action rifle, or a single-shot rifle, or a revolver. What you do need is a semi-automatic firearm that can rapidly fire multiple shots at any wild game animal, or pack of wild game animals, and that can be quickly reloaded using preloaded magazines that you simply insert into the firearm. This type of firearm would maximize your chances of survival and your chances of harvesting wild game for the cook pot. For more information on the advantages and disadvantages of the different types of firearms you should read the firearm chapters in my book **"Grandpappy's Survival Manual for Hard Times."**

Glock Model 22 .40 Caliber S&W Disassembled

You will need a minimum of two firearms in the wilderness. You will need a handgun and you will need a rifle. A shotgun might also be useful in some situations but if you can only carry one long-range firearm at a time then that firearm should be a rifle and not a shotgun.

Your handgun choice should be a semi-automatic pistol. I recommend you purchase a Glock pistol instead of some other brand. I also recommend that you install a LaserMax internal guide rod laser sight inside your Glock. The advantages and disadvantages of laser sights, and instructions for installing a laser sight on a Glock, are in the laser chapter in my book mentioned above.

My first choice in a Glock for full-time wilderness survival would be a Glock Model 21SF (SF = Slim Frame) that fires the 45 caliber bullet. The magazines for this pistol hold 13 rounds and two magazines come with the firearm. You will also need a holster that contains a spare magazine compartment on the front of the holster. This means you would always have a total of 26 rounds of ammunition on you all the time -- 13 rounds in the pistol and 13 rounds in the spare magazine on the front of your holster. A handgun will not kill large wild game animals but it will kill small and medium size wild game animals.

Glock Model 21SF (45 Caliber)

If my hand were a little smaller, or if I were a female, then my second choice in a Glock for wilderness survival would be the Glock Model 22 that fires the 40 caliber bullet. Each magazine holds 15 rounds of ammunition so you would always have 30 rounds of ammunition on your belt. A 40 caliber bullet is less powerful than a 45 caliber bullet and therefore it is not as effective against wild game animals. However, the 40 caliber bullet is very effective against humans and therefore it would be my first choice in a self-defense pistol if I were living in a populated area instead of the wilderness.

Glock Model 22 (40 Caliber)

My choice of a semi-automatic rifle would be based on the following two considerations:

1. the largest wild game animal in the wilderness area where I would be living, and
2. the normal clear visible distance over which I would normally be shooting.

You will need to answer both of these questions based on the exact wilderness area that you have selected. However, the following general comments may assist you in making this decision.

You should consult and obey the hunting laws in your state. Most states have laws that limit the maximum magazine capacity of a hunting rifle to five rounds. This would mean that you may need to purchase a special limited capacity magazine for your rifle, or you will need to disassemble one of your magazines and install a block (wood, metal, or plastic) in the bottom part of the magazine so that no more than five rounds can be inserted into the magazine. Then reassemble the magazine and verify that it works correctly.

You should mount an average or better quality scope on your rifle using see-through scope mounts. The see-through scope mounts will allow you to use your scope for long distance shots or your iron sights for relatively close shots. The scope mounts should have two mounting screws on each side of each mount that holds the scope to the scope mount. This would be a total of four mounting screws on the right side of the scope and four mounting screws on the left side of the scope for a total of eight mounting screws that secure the scope to the scope mount. Instructions for zeroing a rifle scope are in my book that was mentioned previously in this chapter.

If I were living in a wilderness area where there were wild game animals bigger than a deer, or if my average shooting distance would normally be more than 240 yards, then I would select a semi-automatic rifle that could shoot the 308 caliber bullet, such as an AR-10 or a Springfield Armory M1A. The maximum magazine capacity of an AR-10 is 20 rounds, and the maximum capacity of a Springfield M1A is 20. The 308 bullet has 800 foot pounds of kinetic

AR-10 with Stainless Steel Barrel, Scope, and Bipod

energy at a distance of 600 yards so it would normally be lethal out to that range. However, the trajectory and wind drift of a 308 bullet is more challenging to predict at shooting distances beyond 260 yards. I mention this because most of us are average shooters and we can usually hit somewhere inside an 8-inch circle at 200 yards or less. But as we move further away from the target our shot placement circle begins to increase significantly and the chance of our hitting a wild game animal in its "kill zone" rapidly approaches zero. And please don't believe the myth that if you just practice more then you will eventually become an expert marksman. Only about 2 or 3 percent of us are natural sharpshooters. The military knows this and when they find those 2 or 3 men out of 100 then they offer to train them as long distance snipers. But the military also knows that it is a waste of their time and money to try to train someone who is not a natural born sharpshooter to become an expert marksman. This is the same basic concept as trying to learn how to become a professional artist or how to become a professional baseball player. You can practice for years and years but unless you have some natural born artistic talent then you will never become a great artist, and unless you have some natural born athletic talent then you will never become a professional baseball player on a Major League baseball team. Most people understand this concept when it comes to the arts or to sports but they fail to apply it to shooting and they believe the myth that all you have to do is practice more and you will become an expert shot. This is simply not true. After you reach your natural peak shooting performance then additional practice is not going to result in a noticeable improvement in your shooting skills. Therefore the primary reason I would select the 308 caliber rifle is because it will kill larger wild game animals and not because of its maximum lethal shooting distance (unless I was a natural born expert marksman which I am not).

If a deer were the largest wild game animal in my area, and if most of my shooting would be done at distances of 200 yards or less, then I would select a semi-automatic AK-47 that shoots the 7.62x39 caliber ammunition. The AK-47 has magazines that will carry 20, 30, and 40 rounds. My experience has been that the 30 round magazines are usually more reliable than either the 20 or the 40 round magazines so I recommend the 30 round magazines. The 7.62x39 caliber bullet has 800 foot-pounds of kinetic energy at 240 yards and its bullet trajectory will remain within approximately plus or minus 1.5 inches from the front end of the barrel out to a distance of about 175 yards. This makes this bullet an extremely accurate and deadly round for most hunting situations in forested areas on game animals of deer size or smaller. If you can afford it then I recommend an AK-47 with a folding rear stock instead of a wood stock, and a black plastic front hand grip instead of a wood front hand grip. The folding rear stock should fold over against the side of the rifle.

AK-47 GP WASR-10 (Semiautomatic 7.62x39)

The following two firearms are not critical but they could prove to be very useful:

22 Caliber Semi-Automatic Rifle: My first choice in a 22 rifle would be the Ruger Model 10/22 with detachable 10-round magazine. If you can afford it then purchase a 10/22 with a stainless steel barrel rifle. My second choice in a 22 rifle would be the very affordable Savage Arms Model 64F semi-automatic rifle with a detachable 10-round magazine. Regardless of your rifle choice buy some extra magazines. Install a sling and a

Ruger Model 10/22 (Semiautomatic 22LR)

Savage Arms Model 64F (Semiautomatic 22LR)

reasonable quality scope ($30 to $60) on your 22LR rifle using see-through scope mounts. If you can afford it then purchase 10,000 rounds of 22LR ammunition for your rifle.

Beeman Air Rifle 22 caliber and 177 caliber Combo or equivalent (silent harvesting of very small game). You should also purchase 5,000 hollow point or pointed air rifle pellets, preferably in 22 caliber.

Ammunition: You should have a minimum of 3,000 rounds of ammunition for your hunting rifle and a minimum of 1,000 rounds of ammunition for your semi-automatic pistol. The ammunition you take with you into the wilderness may be all the ammunition your group will have for the rest of your life.

You will also need gun cleaning kits, a gunsmith screwdriver set, patches, solvent, and lubricant.

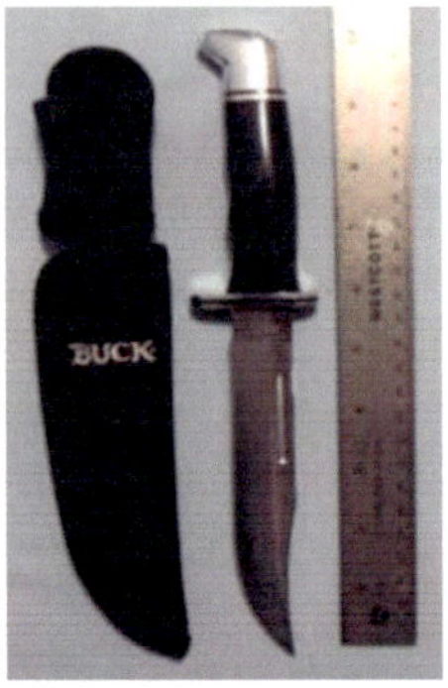

Buck Model 119

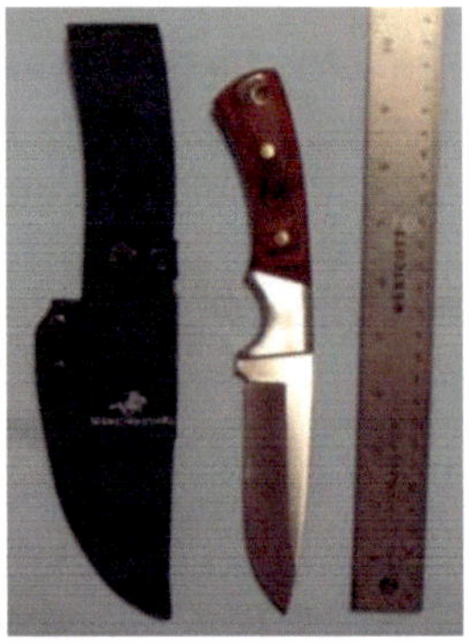

Winchester Basic

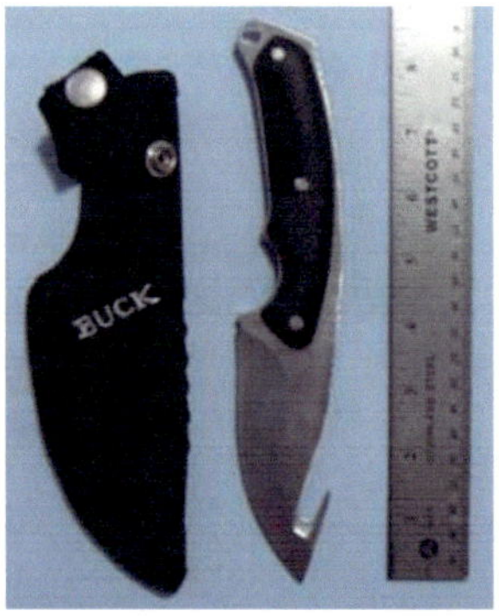

Buck with Gut Hook

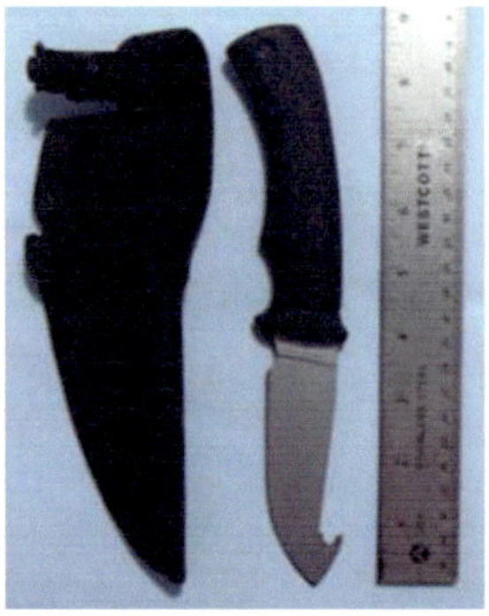

Gerber with Gut Hook

Knives

When you are living in the wilderness the first thing you should do each morning after you get dressed is to strap on your special equipment belt. That belt should contain your pistol inside its holster and a fixed blade hunting knife in its sheath. The last thing you should do each day is remove your equipment belt and put it close to your bed before you go to sleep. In the wilderness you may need your pistol or your knife immediately if you are unexpectedly confronted by a predatory animal that has decided that it is going to try to kill you and eat you, or if you are unexpectedly approached by a group of strangers and you do not know what their intentions might be. Having your pistol and your knife always at hand at your waist will maximize your chances for long-term survival in the wilderness.

My first choice of a hunting knife for a wilderness adventure would be a fixed blade hunting knife in a belt sheath. That hunting knife should not have a gut hook. The reason is because if I needed to use that knife in an unexpected biting and clawing fight with a predatory animal then I would want a knife that could penetrate the wild animal's body quickly and deeply many times in rapid succession and I would not want to take a chance that the gut hook might catch on a bone or sinew and be difficult to withdraw from the body of the animal while the wild animal continues to claw and bite me.

My first choice in a fixed blade hunting knife would be a Buck Model 119 in a black belt sheath for about $40. If money were limited then my second choice would be a Winchester basic hunting knife in a belt sheath for about $16. Either one of these two knives could be used for emergency self-defense against a wild game animal, or against an unethical person who was attacking you. It could also be used to skin a wild game animal or to clean a fish.

However, to skin a wild game animal I would prefer a fixed blade hunting knife with a gut hook, such as a Buck knife or a Gerber knife as illustrated in the above pictures. If I were on a hunting field trip then I would carry my primary fixed blade knife on my equipment belt. I would carry my gut hook knife in my backpack along with my Latex rubber skinning gloves (kitchen gloves) and at least 20 empty one-gallon or two-gallon plastic zipper bags for transporting fresh meat back to my shelter.

Chapter Seven

Water for Drinking and for Personal Hygiene

Being able to continually replenish your drinking water is absolutely critical to your survival in the wilderness. A person who does not drink any water or other fluids for three days will die of dehydration and delirium. Therefore a continuous supply of fresh clean drinking water should be your first critical priority in the wilderness.

A reasonable supply of drinking water is also one of those items you must take with you into the wilderness. You should have enough drinking water to last each member of your group until the next time it rains, or until you can locate a fresh water spring head. The optimal solution would be to purchase between one to three cases of drinking water per person and then take those cases of drinking water with you into the wilderness. I recommend the 20-ounce plastic water bottles that are sold either 24 or 30 per flat case. Invest in water that is inside strong thick-plastic water bottles instead of the flimsy thin-plastic water bottles. The reason is because you can save the empty plastic water bottles and then refill them with fresh water. This will allow you to reuse those same plastic bottles again and again and again for a very long time. And since they are sold on a cardboard flat you can stack them neatly inside your shelter, one case on top of another, and they will not take up a lot of space. Each person in your group should write his or her initials on his or her own water bottles using a black permanent felt tip marker in order to avoid the transfer of germs between the people in your group.

Drinking Water

Ground Water: A lake, pond, or creek does not contain pure clean drinking water. Any water that is on the surface of the earth is polluted water. That water contains harmful microorganisms that will either make you extremely sick or kill you. All types of animals, reptiles, birds, and bugs live in the wilderness. Every single one of these creatures periodically relieves itself of its bodily wastes somewhere on the ground. And each time it rains those waste products are washed down into the nearest creek, pond, or lake. Therefore if you drink creek, pond, or lake water then you will be drinking water that contains a multitude of tiny microscopic organisms that will multiply inside your body and eventually kill you.

Rainwater: If you can catch rainwater before it makes contact with the ground, or with any trees or rocks, then that rainwater will not be contaminated. It will not contain the tiny pathogens that are present in creek, pond, and lake water. However, if the air contains smog, volcanic ash, or nuclear particles, then even rainwater will not be fit for human consumption until after you have filtered it.

Inches of Rain: Before you move into the wilderness you should find out what the historical average rainfall has been for the county in which you plan to live. You will need to know the average rainfall for each month of the year. In some areas the rainfall is relatively stable all year but in other areas there is a rainy season and a dry season. You need to know this information about your area before you move into it. This will help you to plan when you will harvest the rainwater and how much rainwater you will need to store in order to survive the dry season. You may discover that the dry season is so long that it would not be possible to store enough rainwater to last for that length of time and therefore your first choice of a wilderness area will simply not be feasible because you would die of thirst. However, if the average monthly rainfall is adequate then the wilderness area you have selected might be suitable for long-term survival.

How to Collect Rainwater

To capture rainwater you will need several heavy-duty tarps. These heavy-duty tarps should be 10 mil thick. The size of the tarps should be somewhere between 9-feet by 10-feet up to approximately 12-feet by 20-feet. You will also need some strong nylon or polypropylene cord or twine to secure the grommets in the outside edges of each tarp to some nearby objects (such as trees) so the tarp will hang down in a "U" or "V" shaped pattern from side to side and with the rear of the tarp about one-foot higher than the front of the tarp. Place a large clean empty plastic tote container at the front end of the tarp at the center of the "V" to catch the rainwater as it pours out the front end of the tarp. (Note: One gallon of water = 231 cubic inches of water.)

As a practical example, if it rains one-inch and you have the tarp arranged in a "U" or "V" shape then:

1. A single 9-foot by 10-foot tarp will collect approximately 50 gallons of rainwater.
2. A single 10-foot by 12-foot tarp will collect approximately 70 gallons of rainwater.
3. A single 12-foot by 20-foot tarp will collect approximately 140 gallons of rainwater.

For planning purposes you will need approximately one-gallon of water per day per person. This is enough water for drinking and for brushing your teeth. It is not enough for cooking, or bathing, or washing the dishes, or doing the laundry. These activities will require significantly more water. You should have enough drinking water for each member of your group to last at least 30 days. If your area has a dry season with very little or no rain for several weeks or months, then you will need to have enough water to last the entire dry season. You will need to do the math and determine how many tarps you will need to satisfy your water requirements so you do not die of thirst during the dry weather.

Water Filters

Rainwater is safe to drink if you can catch it before it comes in contact with anything else. In other words, if you stand outside with your mouth open and the rain falls straight down from a cloud in the sky into your mouth then you could swallow that water and not worry about getting sick (assuming the air was not full of smog or something else).

However, if the water makes contact with anything else, such as the leaves on a tree, then you will need to process that water before you drink it. There are also several other factors that may force you to process your rainwater before you can drink it. Some examples would be:

1. A limited nuclear war, or a nuclear power plant meltdown, anywhere on the face of the earth. The tiny nuclear particles can travel extraordinary distances after they have entered the upper atmosphere. If you are filtering out radioactive particles then position your water filter system far enough away from your living quarters so the tiny radioactive particles that will get trapped inside the water filter cannot slowly poison the members of your group. Radiation only kills living organisms. It does not kill non-living things such as water or canned food. However, if you ingest water or food that contains radioactive particles then you will get sick and eventually die. But if you remove the radioactive particles from the water then the water will be safe to drink. In other words, after you have removed the radioactive particles from the water then the water could be brought into your shelter but the water filter that trapped the radioactive particles should not be brought into your shelter.
2. A volcanic explosion that pumps huge amounts of volcanic debris into the upper atmosphere.
3. A small meteor that hits the earth and sends huge amounts of debris into the upper atmosphere.

If any of the above events occur then it would not be safe to drink rainwater until after you have filtered it. You have two different options for filtering your rainwater and you will need both of them.

1. **An AquaRain Model 400 Gravity Water Filter System** (a stainless steel water storage tank and four filters). It uses gravity to filter the water and it comes with four 0.3 micron filters. Each filter has a maximum rated life of 10,000 gallons of water under good conditions. It weighs about 10 pounds and it has exterior dimensions of 10.5" diameter and 22" tall. It can be used with one, two, three, or all four filters installed. One filter will yield 6 gallons of water per day. If you install and use all four filters at the same time it will process 24 gallons of water per day. The filters need to be periodically removed and boiled for 15 minutes and then allowed to cool. Before putting water into the upper water filter compartment you should pour the water through a clean cloth to remove any large impurities. Then boil the water for one-minute to kill all the life-threatening pathogens in the water. Wait for the water to cool and then put the water into the upper water container. The filters will then remove any particles that are 0.3 microns or larger in size.

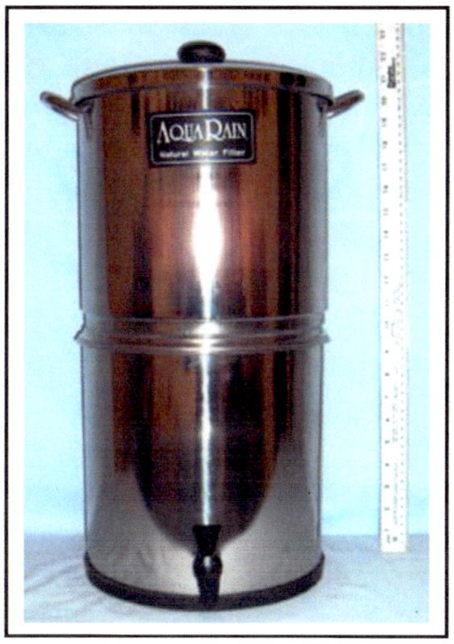

2. **A Swiss Katadyn Pocket Water Filter**. It is portable. It comes with one 0.2 micron filter element that will yield one-quart of drinking water per minute. The filter has a maximum rated life of 13,000 gallons of water under good conditions. Its disadvantages are that it must be hand pumped, and if its one filter is damaged in any way then it makes the unit of questionable value. However, if you will not always be returning to your shelter every evening then you will need this filter in your backpack. (Note: There are several other cheaper Katadyn water filters available, such as the Hiker and the Guide, but they are not designed for a long-term survival situation in the middle of a wilderness area.)

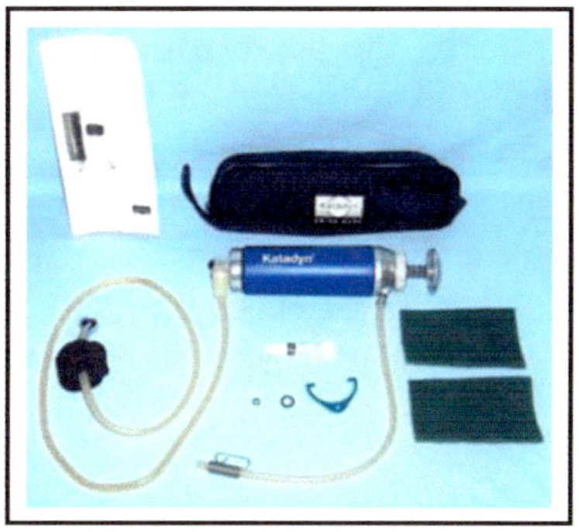

Where to Collect Rainwater: You should consider collecting your rainwater as close as possible to your shelter. If possible collect your rainwater from a slightly higher elevation than your shelter so you can use gravity to transfer the water to your shelter.

Where to Store your Rainwater: If possible you should store your rainwater inside your shelter or as close as possible to your shelter.

How to Store Rainwater: One cheap way to store a lot of water is to use 5-gallon food grade plastic buckets or clean empty large heavy-duty plastic tote containers. Please remember that water is very heavy so please don't try to move a heavy plastic tote full of water. Five gallons of water inside a 5-gallon plastic bucket weighs approximately 42 pounds. Do not injure your spinal column trying to move too much water at one time. If you must move your water then only move a few gallons at a time. This may require several trips back and forth from your water collection site to your water storage site. But you should remember that there are no emergency first aid services in the wilderness and if you injure your back then your chances for long-term survival in the wilderness will quickly approach zero.

Or you could store your extra water in clean empty two-liter soft drink plastic bottles with screw on caps. Or you could refill empty plastic water bottles that you saved for this purpose. These would be the 16-ounce to 20-ounce clear plastic water bottles that are sold by the case.

Spring Water

When it rains the water does two things. Some of it travels along the top surface of the ground and ends up in creeks and streams and lakes. But a lot of it soaks into the earth and some of it makes its way down to the natural water table in the area. The earth is an excellent water filter. If the water table is

100 feet or more beneath the surface of the earth then the water there is usually safe for human consumption without any treatment. That's why people in the country drill deep wells -- they want clean water that doesn't need treatment.

The next question is, how does a person get to that deep water out in the middle of the woods? Do not waste your time trying to dig for it. It isn't worth the effort.

The best way is to find a spring head. Let me explain. Because of gravity water seeks the lowest possible level. You can usually find a stream or creek at the bottom of most hills or mountains. When you find one then begin walking upstream. Every now and then you will find a tiny feeder stream flowing into the main stream. The main stream will continue along the foot of the hill. Follow the tiny feeder stream up the side of the hill. Sooner or later you will eventually come to its source. It will just emerge from the ground. This is called the "head of the spring." The underground water table in this area is very close to the surface of the earth at this one spot and this particular exit point just happened as a result of nature. The water at the head of the spring hasn't had a chance to become polluted with lots of animal waste so it is the cleanest water you will find in the woods.

Spring Head inside White Oval. Water Flows in Direction of Arrow.

The head of a spring is normally a reasonable distance up the side of a hill or mountain. If there is any level ground near the spring head then this is usually an excellent place to establish your permanent shelter. You avoid both the peak and bottom of the hill (high winds and flooding waters), and you are near a really clean reliable source of good water.

A Manmade Waterhole

Another option for collecting and storing rainwater is to dig a hole about one-foot deep and about three-feet shorter than one of your heavy-duty tarps. This would be a manmade waterhole or a manmade pool.

When you dig the hole you should transfer the dirt from inside the hole to the outside edges of the hole so the fresh dirt forms a slightly elevated mound around the entire outside border of your new waterhole. This short mound or hill all the way around your waterhole will prevent rainwater from flowing across the ground and into your waterhole each time it rains. Remember that the ground contains germs and animal droppings and you do not want that stuff inside your waterhole.

The bottom of your waterhole should be flat and it should be lined with a heavy-duty tarp that is approximately three-feet longer and three-feet wider than the bottom of your hole. This would provide one-foot of tarp to place against each side of the pool, plus an additional six-inches to place on the ground around the entire outside edge of your pool. You could use nylon or polypropylene twine to tie each grommet hole on the outside edge of the tarp to a wooden stake that you hammer into the ground below each grommet hole. The tarp inside the waterhole would serve the same basic principle as a swimming pool liner.

The pool should then be completely covered with another tarp to keep leaves and sticks and bird droppings out of the water, and to prevent sunlight from hitting the water. The sun helps all types of germs to multiply. It is more difficult for germs to multiply in a cool area in the dark. The tarp covering would also minimize water loss due to evaporation because the water would condense on the top tarp and drop back down into the pool. You could support the tarp over the pool using poles that you cut to support the tarp. You would need to build a framework that could support the weight of the tarp and keep it from hanging down into the water. Cover the top tarp with a few leaves and small sticks to help conceal the location of your waterhole.

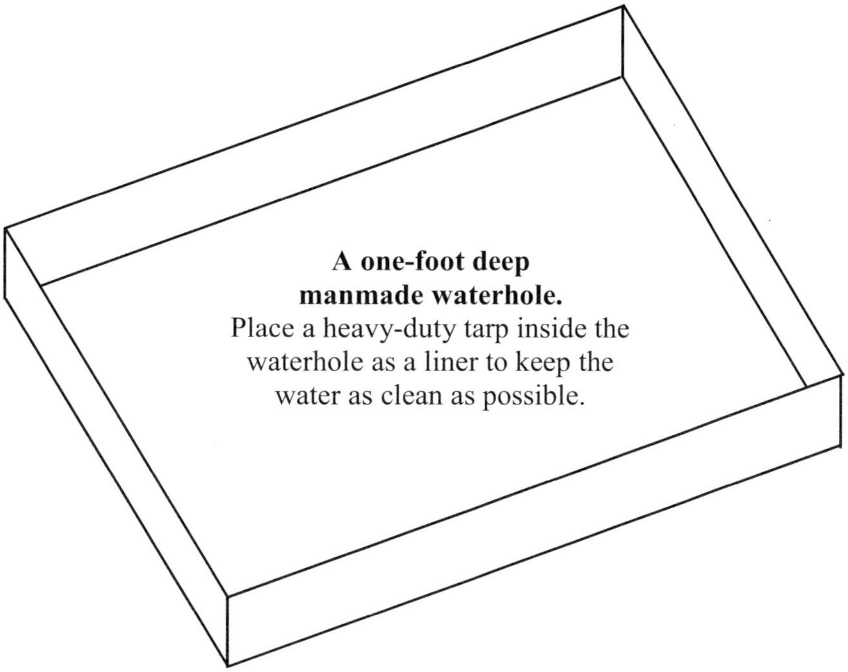

However, a multitude of insects will still find your water and get into it so you will need to filter your water before you use it. Wild animals will also be able to smell the water and this may attract them to your underground waterhole. If you place a network of steel traps and snares in the paths and game trails that lead towards your waterhole then this could easily result in some fresh meat on a regular basis for your dining pleasure.

You will also need to secure several other tarps above and around the outside edges of your waterhole in an "U" or "V" shaped pattern as explained earlier in this chapter. These tarps will catch and direct the rainwater into your waterhole. The bottom of the "U" or "V" should be directly above the outside corners of your waterhole where you have carefully folded back the top tarp just a little in order to allow the flow of rainwater into your pool. You will not be able to leave the rainwater tarps in position all the time because they could be too easily seen by anyone who happens by, and they will also gradually collect bird droppings. Bird droppings can be fatal if they get into your drinking water. Therefore just before it rains, or immediately after it starts raining, you will need to secure your "U" or "V" tarps into position. And when it stops raining you will need to take them down again. You will also need to fold the outside corners of the top pool tarp back into position to cover the pool of water.

It would probably be better to construct two or three smaller waterholes instead of one huge waterhole. This would allow you to completely use all the water in one hole and then remove the tarp liner and scrub any algae and dirt off the tarp liner using some soap and water. Then you could rinse off the tarp and reinstall the tarp in the bottom of your hole and fill it up the next time it rains.

For example,

1. A 5-foot wide by 6-foot long hole that is one-foot deep will store about 190 gallons of water.
2. A 6-foot wide by 8-foot long hole that is one-foot deep will store about 300 gallons of water.
3. A 7-foot wide by 9-foot long hole that is one-foot deep will store about 390 gallons of water.
4. An 8-foot wide by 10-foot long hole that is one-foot deep will store about 500 gallons of water.

Water Transfer

You can transfer water from your "waterhole" or from your "natural spring head" to your shelter using five-gallon food grade plastic buckets. Or you could install an ordinary garden hose between your "waterhole or spring head" and your shelter. Put one end of the hose in the bottom of your waterhole or spring head and attach a standard water hose nozzle or a standard faucet onto the other end of your water hose. Close the nozzle or faucet end of your hose. Then use a bucket or a cook pot to gradually fill your entire water hose with water by pouring water into the open end of the hose. The water will force the air out of the hose and the hose will gradually fill with water. When the hose is completely full of water then put the open end of the hose in the bottom of your waterhole or spring head. If you have a small piece of plastic screen then secure it over the end of the hose to filter out any large foreign particles, or insects, that might get into your water. When you open the nozzle or faucet end of your hose the water will flow through the hose to your shelter location. *However, for this siphoning action to work the nozzle or faucet end of your water hose must be lower in elevation than the bottom of your waterhole.* After you have a working water transfer system then you should bury the water hose about six-inches below ground the entire distance between your waterhole and your shelter to conceal it.

You could also transfer the water using plastic hot water pipes and the appropriate type of pipe glue. Hot water pipes are stronger than cold water pipes and therefore they would be better suited to a wilderness application. However, plastic water pipes may be more challenging to transport into a wilderness area when compared to a few ordinary garden hoses. You will need to make your own decision about this issue as it relates to your specific application.

If you will utilize either a water hose or water pipes to transfer your water from higher ground to lower ground using gravity then you will be able to eliminate the work of carrying water by hand from your water collection area to your shelter. This will not only save you a considerable amount of time but it will also allow you to utilize your available energy on a more productive activity.

Snow or Ice

Bring the snow or ice inside your shelter and allow it to melt into water. You will be surprised at the small quantity of water you will get from a big bucket of snow. Put the pot of snow in the sun in front of a window and wait for it to melt. If necessary you can put the snow inside a small or medium size container. Then put the capped container inside your clothing (but not next to your skin) and your body heat will gradually melt the snow into water. If you are concerned about the cleanliness of the snow then you can boil the resulting water for 1 or 2 minutes before you filter it. Never melt snow or ice inside your mouth. It will not relieve your thirst and you will dehydrate more quickly. Always melt the snow or ice first and then filter it and then drink the water.

Boiling Water

Boiling is the BEST method for killing all the pathogens in the water. Even the most expensive commercially available water filter can't make that claim.

Water boils at approximately 212°F (100°C) at sea level at a barometric pressure of approximately 30.3 inches of mercury. The boiling point of water decreases as the barometric pressure decreases. Therefore the changing barometric pressure in your area has as much impact on the boiling point of water as the altitude at your location. At 8,000 feet above sea level and at a barometric pressure of approximately 27 inches of mercury, water boils at approximately 189.4°F (87.4°C).

All pathogens die rapidly at 185°F (or 85°C). Some pathogens die at lower temperatures. By the time the water has reached its boiling point (even at low barometric pressures), all the pathogens in the water are already neutralized.

Therefore bring your water to a boil and let it hard boil for one-minute. Boiling for more time doesn't help, and it results in more water being lost as steam, and it makes the water taste flatter. All the pathogens are already dead by the time the water reaches its boiling point. There is no benefit to a pathogen being "more" dead.

Wait patiently for the water to gradually cool to a comfortable drinking temperature.

However, boiling will only kill the microscopic pathogens in the water. Boiling will not remove any dissolved chemicals in the water, or the salt from salt water, or any nuclear radioactive particles. If any of these issues are a concern then you will absolutely need to filter your water after you have boiled it and before you drink it (or you will need to distill your water using a solar still).

Some Options for Boiling Water

Stainless Steel Coffee Pot (1 to 2 quarts): Relatively fast and it yields the approximate amount of water that a family will drink at one meal. It has a convenient handle that makes it easy to pour the water into cups or into another container. Obviously it could also be used to brew fresh coffee.

Fill the left Basket with Ground Coffee and then put it inside the Coffee Pot on the right.

Six Quart Cook Pot with a Lid and a Handle: The lid will keep air-borne particles out of the water. The handle will allow you to easily pour the water into another container.

Five Gallon Stainless Steel Cook Pot with a Lid: It takes longer to bring five gallons of water to a boil. However this is enough water to last the average family an entire day for all their drinking and personal hygiene purposes. You will need to transfer the water to another container using a separate sterile pot with a handle.

Solar Still

If you need to remove the salt from sea water, or if you need to remove radioactive particles from your water, then one method that would work in both of these situations would be to construct a "solar still" following the detailed instructions in the water chapter of my book **"Grandpappy's Survival Manual for Hard Times."** After digging your solar still you should place your contaminated water in several buckets that you plant on the inside slopes of your solar still and then wait for evaporation to transfer the water from the buckets to the underside of the tarp. That water will form droplets on the underside of the tarp and then flow down the underside of the tarp and the water droplets will fall into your clean water container in the bottom center of your solar still. However, the other buckets will still contain the sea salt and the radioactive particles because salt and radioactive particles do not evaporate.

Water for Personal Hygiene

You will need water to wash your dishes and your laundry, and for personal hygiene purposes. For safety purposes you should filter the rainwater you drink, brush your teeth with, and cook with. However, you will usually not need to filter rainwater for laundry and bathing.

Shower: You should have a portable "solar-heated camp shower bag" that you can put in the sun each morning so the sun can heat 4 or 5 gallons of water to a very comfortable warm temperature. You will then be able to take a nice warm shower at the end of each workday. If you have two solar shower bags then you will also have some warm water to wash your dishes.

Outdoor Toilet: You will not have a composting chemical toilet. Instead you should have a supply of toilet tissue and a portable camping toilet stool that unfolds (no plastic poop bag). You will place that toilet stool over a small hole that you dig in the ground. After you have a bowel movement you will fold the toilet up, cover the small hole with some dirt, and forget about it. You will then immediately dig another new shallow hole beside the previous hole that you just filled up so the new hole will be ready the next time you need to use the toilet.

Indoor Toilet: You should also have an indoor toilet for use in the middle of the night and during really cold winter weather. This will consist of a "bucket toilet seat" that can be attached to a standard five-gallon or six-gallon plastic bucket with a tight fitting lid. You can put the lid back on the bucket after you use the toilet to prevent the odor from smelling up your shelter. You will periodically need to empty the contents of this bucket into a hole outside your shelter and then cover your waste materials with some dirt. Or you could empty the toilet bucket into a compost pile and convert the human waste into a safe garden compost fertilizer following the instructions in the compost chapter of my book **"Grandpappy's Survival Manual for Hard Times."**

The Effort Required to Collect and Store Water and to Dispose of Human Waste

Collecting, moving, and storing your water will take some time and hard work, and that is work you did not have to do when you lived in the civilized world. Therefore when you first begin this activity you will resent the fact that you now have to do it because you have probably become accustomed to simply turning on the faucet and having hot and cold running water instantly on demand. You need to be aware of this from the very beginning so you will be intellectually prepared to deal with all the hard work that will be involved in providing for your daily water requirements.

Chapter Eight

Food Recommendations and How to Replenish Your Food Supplies from Nature

The four basic necessities of life are air, water, food, and shelter. Water was discussed in detail in the previous chapter. This chapter will discuss food.

If you have enough water to drink, enough food to eat, and a safe shelter to sleep in at night (and to relax in when it rains), then you can survive for a very, very long time in the wilderness. But if you don't have water, or food, or a safe shelter then your life expectancy in the wilderness will be relatively short.

One of the most common mistakes that people make is to assume that they can enter the wilderness and find enough edible wild plants, fish, and wild game to keep them fed for a very long time. This is an invalid assumption that almost any hunter will verify if you ask him. Hunting and fishing is a time consuming activity that does not always result in a harvest of fish or wild game.

And only a very limited number of edible wild plants grow in each wilderness area. When you read a survival manual and you see pictures of over a hundred different edible wild plants then it easy to believe that the wilderness is a huge grocery store just waiting for you to harvest it. This is a pleasant daydream but it is just an illusion. Only a few edible wild plants grow in each wilderness area and most of those plants are not in season all year. And when they are in season there is usually only a very limited supply of them and the forest wildlife will eat them as soon as they are mature enough to eat. Wild berries are a good example. They are in season for about three weeks during warm weather and as the berries gradually ripen they are stripped from their vines by birds, small animals, and bears.

Therefore if you wish to survive for several years in a hostile wilderness environment then you will need to take enough food with you to last at least two years. During this time you can gradually become familiar with the edible wild plants that grow in the region where you live, and you can gain some practical experience using steel traps and snares to catch small and medium size wild game animals, and you can invest some time in hunting the larger wild game animals, and you can learn how to successfully use a gillnet to harvest fresh fish. But each of these skills takes time to learn through trial and error and you will need food to eat while you are learning these basic survival skills. You will also be planting some basic garden vegetables and learning how to preserve those vegetables for consumption during the cold weather months. Finally, you will plant some fruit trees but those fruit trees will not yield any fruit for at least three years and maybe even longer.

Food Recommendations

Twelve-month supply of basic food items (three-years would be better).
200 Mainstay Brand Marine Lifeboat Ration Food Bars.
200 pounds of wheat berries (red and white) sealed in mylar bags inside six-gallon food pails.
One hand-cranked wheat berry grinder such as the Family Grain Mill.
50 pounds of pure salt (Morton Canning and Pickling Salt with no additives).
10 pounds of Morton Lite iodized salt.
20 pounds of pepper or whole peppercorns (and a pepper grinder).
100 Foil-Packed Brown Gravy Dry Mixes.
100 Foil-Packed White Gravy Dry Mixes.
100 Foil-Packed Taco Seasoning Dry Mixes.
100 Foil-Packed Chili Seasoning Dry Mixes.
100 Foil-Packed Sloppy Joes Seasoning Dry Mixes.
100 Foil-Packed Beef Stew Seasoning Dry Mixes.
100 Jars of Seasoned Meat Tenderizer (5th Season 5.5 ounce jars).
500 multi-vitamin tablets (take one table every other day or one tablet every third day).
100 pounds of hard candy, miniature tootsie rolls, and/or caramels.

Stainless steel heavy-duty hand-operated can opener.
Two 48-quart (or larger) heavy-duty premium quality ice chests.
100 one-quart heavy-duty plastic food storage containers with snap on lids (the nesting type).
300 Ziploc heavy-duty one-gallon **freezer** bags (may be "reused" several times for future food storage).
Heavy-Duty 18-inch wide aluminum foil (use to build a solar oven).
One box of 5 Reynolds Oven Bags (use to build a solar oven).
Assortment of heirloom vegetable seeds (at least $50 worth).
Fruit trees, blueberry bushes, grape vines (seedlings or three-year old plants).
Fishing tackle gear, lots of fish hooks, collapsible fishing rod, and two gillnets.
2,000 yards of quality fishing line.
Conibear traps, pan traps, and professional quality snares.
100 yards 14-gauge wire.
12 Pair of quality Latex gloves (one size larger than you normally buy -- clean fish and wild game).
Hand-cranked stainless steel meat grinder for the occasional tough wild game meat.

Marine Food Bars: Marine food bars contain an assortment of vitamins, they are not thirst provoking, they have a five-year shelf life, they can tolerate temperature extremes from -40°F to +300°F (-40°C to +149°C), they have a pleasant taste (in my opinion but my youngest son disagrees), and you can buy them at some marine supply stores, some Army/Navy stores, some camping stores, or over the internet. They can also tolerate temperature extremes from -40°F to +300°F. If they are stored in a more suitable temperature controlled environment then their shelf life could easily exceed five-years.

There are two major brands of Marine Food Bars: Datrex and Mainstay. They are sold in different package sizes but the 3600-calorie size is the best buy. Each 3600-calorie package contains nine 400-calorie food bars. The Mainstay Food Bar intentionally has less protein in order to minimize the amount of water required for proper digestion because that is necessary in a marine survival scenario. The Mainstay brand is the better value in my opinion. The following table compares the two brands:

Description	**Datrex**	**Mainstay**
Flavor	Mild Coconut	Vanilla/Lemon
Calories	400	400
Protein	6 g or 14%	3 g or 7%
Cholesterol	2 g	0 g
Carbohydrates	43 g	46 g
Vitamin A	2 %	50 %
Vitamin B6	0 %	90 %
Vitamin B12	0 %	20 %
Vitamin C	2 %	60 %
Vitamin D	0 %	50 %
Vitamin E	0 %	25 %
Calcium	0 %	50 %
Folic Acid	0 %	30 %
Iron	0 %	10 %
Magnesium	0 %	30 %
Niacin	8 %	30 %
Pantothenic Acid	0 %	100 %
Phosphorous	0 %	40 %
Riboflavin	10 %	25 %
Thiamine	16 %	15 %

Food: When you are making your food purchases you should select a reasonable variety to prevent appetite fatigue. A recommended one-year emergency food supply is in my book **"Grandpappy's Survival Manual for Hard Times."** However, when you are living in the wilderness it is helpful if you have one-can or one-pot complete meals in order to simplify the cooking or heating process, and to minimize future cleanup operations. Therefore you should carefully consider one-can meals that you actually enjoy eating. Some examples might be: Beef Stew, Chili with Beans, Spaghetti with Meat Balls, Beef Ravioli, Roast Beef Hash, and Campbell's Chunky Soups. Before you purchase a huge inventory of any food item you should purchase one can and take it home and eat it. If you find the taste acceptable then purchase additional quantities. But if you find it disgusting then don't buy any more of that brand or that particular food item. If you have an adequate food supply then your mind can focus on other more critical issues. But if you don't have food then you will not be able to think about anything except your lack of food.

Dry Mixes: The recommended food list in this chapter includes 600 foil sealed dry powder mixes (brown gravy, white gravy, taco seasoning, chili seasoning, Sloppy Joes seasoning, and beef stew seasoning). In most cases a single seasoning package contains enough mix to prepare two or three meals so don't use an entire package at one time. Only prepare one-half or one-third of the package depending on your current need. You would be astonished at how good wild game and wild plants will taste if they are covered with a seasoning you are already familiar with. If you do not enjoy the taste of one of the recommended dry seasoning mixes then do not buy it. Also feel free to substitute any other dry seasoning mix that you prefer. However, I suggest you buy the generic brand or the store brand whenever possible to get a better value for your money. In most cases these generic packages cost about 50-cents each. If you invest one-dollar or more in a seasoning mix package then you would have been better off if you had simply invested that much money in one or two complete canned meals, such as beef stew or chili with beans or canned pasta or Chunky soups.

Seasoned Meat Tenderizer: The reason seasoned meat tenderizer is recommended is because it is really cheap at the current time (50 cents per 5.5 ounce jar at Walmart). This seasoning will make it a lot easier for your entire group to gradually adjust to the flavor of any "wild game meat" you may be able to acquire in the wilderness.

Candy: In addition to nutritious food you should also include a reasonable supply of candy. Select a variety of different types of wrapped hard candy or any candy that will not go bad in a short period of time, and which will not melt if it gets warm. (If possible, seal your freshly bought candy in a vacuum bag using a food vacuum sealer to significantly extend is fresh taste and shelf life.) Having something sweet to eat at the end of a meal gives you something to look forward to even if the meal itself is not that great. When you first arrive in the wilderness you will need to resist the temptation to eat most of your candy during the first few weeks or months you are there. One way to accomplish this is to transfer a one-week supply of candy into a zipper freezer bag and then only eat the candy in that bag during a seven-day period. If you empty your bag before the seven-days are up then do not refill it. For example, every Sunday morning you could fill your zipper bag with your candy ration for that week. If you can see how much candy is still in your bag then it may help you to resist the temptation to eat too much of it at one meal. However, if you simply withdraw the candy each day from your huge candy stockpile then it won't be long before your candy stockpile is completely gone.

Canned Food: If possible, do not eat all your store bought food the first year. Ration your food so it will last at least three years. As soon as possible you should begin to supplement your food with what you can find in nature (fish, wild game, and edible plants). After your new wilderness shelter is completely finished and you have moved into it, then you should select at least one day of the week as your wilderness food day, such as every Wednesday (if it is not raining or snowing). When you get up that morning you decide that you will not eat anything that day unless you can catch it, trap it, shoot it, or find it growing wild. This puts a little bit of pressure on you and it provides an incentive to put forth your best effort. And if you don't find any food then do not eat that day. However, during these first few months the psychological pressure will not be severe because you will know you have a good meal waiting for you the first thing the following morning even you if don't find anything to eat that day. But this will help to prepare you for some serious hard times later after all your store bought food has been consumed. Remember that your long-term objective is to learn how to grow your own food, and find edible berries and nuts, and learn how to fish, trap, and shoot wild game so you can gradually build a sizable winter food supply during the summer and fall.

Therefore you will need to carefully scout the area near your shelter and set a few of your traps to gain some practical experience with trapping, skinning, butchering, and cooking wild game.

Food Replenishment

Fruit: When you first arrive at your new wilderness location you should immediately plant a few grape vines, a few blueberry bushes, and a few fruit trees. Plant them where they will get full sun all day. Full sun and good soil are critical to the success of your new plantings. Water your new plants twice a week. It will be at least three years or longer before you get any fruit from these plantings. You will need to select varieties that do well in your geographical area. The best source of this type of information would be a book on the topic that is available at almost any store that sells fruit trees and grape vines. Resist the temptation to make your decision based on the advice of an employee who works at the store. In most cases the employee will not be an expert in the type of plants you are looking for but he or she will talk and act like an expert and you will not become aware of the bad advice until several years later when you get no fruit from your trees or vines.

You will usually need at least two different varieties for successfully pollination. My book **"Grandpappy's Survival Manual for Hard Times"** contains information on which varieties are compatible for cross pollination. You should plant at least three different plants in the event one plant dies. In other words, plant three apple trees, three pear trees, three plum trees, three blueberry bushes, and three grape vines. If one of the trees or vines dies you should still achieve successful pollination from the remaining two plants. (Note: If fig trees, pecan trees, almond trees, or black walnut trees grow in your area then plant them also.)

Do not buy dwarf fruit trees or semi-dwarf fruit trees. Their roots starve the tree to keep it small. This will result in the tiny fruits dropping off the tree early each spring because they can't get enough nutrients from the tree roots.

Another option would be to grow your fruit trees from the seeds you remove from the fruit you normally eat, such as apples, peaches, pears, and cherries. Complete instructions for growing fruit trees from seed are in my book **"Grandpappy's Survival Manual for Hard Times."** However, that book does not explain how to grow seedlings in small containers and then take those seedlings with you. This may be a more feasible option for someone who is moving into a wilderness area because tiny seedlings do not require the same amount of space as a three or four year old fruit tree. Therefore please allow me to explain how to grow seedlings and then later transfer them to your future wilderness location.

Apple Seeds

You will need some large plastic containers, such as disposable soft drink cups or coffee cups. These would be the large beverage cups you are currently given at a fast food restaurant or gasoline station. You should purchase the largest size beverage possible in order to provide more space for the roots of your tiny fruit tree seedlings. After you drink the beverage take the disposable cup home and carefully wash it out the same way you would your other dishes. Then puncture about six small holes, about one-sixteenth of an inch in diameter and about one-half inch above the bottom of the cup and around the outside of the cup. Then fill the cup with good rich topsoil to within one-half inch of the top of the cup. Then plant one fruit tree seed in the middle of the dirt. Immediately write the name of the seed on the side of the cup with a black permanent felt tip marker, such as "Golden Delicious Apple." Follow the watering and sunshine instructions in the above mentioned book and your seed will gradually grow into a short fruit tree called a "seedling." These seedlings will remain relatively small their first year so you could easily put all your seedling cups into a cardboard box and take your entire future fruit tree orchard with you to your new wilderness home. When you arrive you could transplant your seedlings in a good sunny location and wait for them to grow into a fruit tree but this will usually take another six to eight years. But you will have trimmed one year off the total time by starting your seedlings at your current location prior to moving into the wilderness.

Garden: Plant a very, very small garden the first year. This will give you some first-hand experience with the local growing conditions and the local insects and other pests. Do not plant more than 10% of your **heirloom** vegetable seeds the first year. In addition to growing a few fresh vegetables to eat, you should also learn how to harvest new seeds for planting the following spring. This information is in the gardening section of my book **"Grandpappy's Survival Manual for Hard Times."** Be intellectually prepared to be disappointed the first year you plant a small garden. Think about your first year as a learning experience where you will become aware of the problems in your geographical area so you can avoid most of those problems the next year. If you keep at least 90% of your original heirloom seeds in reserve that first year then you will have some peace of mind during the first winter because you will know that you will be able to plant your seeds the following spring. The second spring you should plant about one-half of your remaining seeds and keep the other half in reserve for the third year. If you always have some heirloom vegetable seeds in reserve then will have hope during the hard winter months.

The best vegetables to plant in a wilderness environment are root vegetables. The edible part of a root vegetable grows below ground and therefore it is invisible unless you know what is growing below the vine or leaves you see above ground. Some common root vegetables that you might wish to consider are potatoes, carrots, beets, turnips, radishes, peanuts, and onions. However, onions can be smelled from a good distance away even though they are growing underground.

Edible Wild Plants: You should learn how to harvest white oak acorns, hickory nuts, and wild berries. These free food items will add some pleasant variety to your meals. Information on how to harvest, process, and store white oak acorns is in my book **"Grandpappy's Recipes for Hard Times."** If other edible wild foods grow abundantly in your area then you should become familiar with them also. Resist the temptation to eat any wild mushrooms even if they are "safe." Mushrooms contain almost no nutritional value and a single mushroom mistake could quickly cost you your life.

Wild Game: You should learn how to catch wild game using professional quality snares and steel traps. You will probably not be very successful the first year but the food supply you bring with you should be adequate during this learning period. You should do much better the second year. By the third year you should not have to worry about meat. In the spring, summer, and fall you will be able to enjoy a steady supply of fresh meat, and in the winter you will have dried meat jerky. Instructions for processing and cooking a variety of different wild game animals are in my book **"Grandpappy's Recipes for Hard Times."**

Animal Fat: You must cut off and remove the fat from any wild game animal that you butcher. You should then render that fat by melting it in a very small amount of pure water. If you do not render the fat very soon after the death of the wild game animal then the fat will become rancid and you will have to throw it away. However, properly rendered animal fat may be used to make pemmican, candles, and soap. My homemade soap recipe is in my book **"Grandpappy's Survival Manual for Hard Times."** In my opinion the best use for animal fat in the wilderness is to use it as one of the ingredients in pemmican, a native American Indian survival food. My meat jerky recipe and my pemmican recipe are in my book **"Grandpappy's Recipes for Hard Times."**

Fresh Fish: You should learn how to use a gillnet to easily capture fresh fish for the frying pan. There is nothing wrong with fishing with a conventional fishing rod and a reel but this type of fishing takes time. There will normally be some many things that you need to do each day that you will not have the time to invest in a leisurely fishing expedition. This is where a gillnet will be of tremendous value to you. After you place the net into its position in a lake or stream you can leave it alone and then come back 12-hours later and remove any fresh fish that have been caught in the gillnet. Complete instructions for constructing and using a gillnet are in my book **"Grandpappy's Survival Manual for Hard Times."**

Smoke House: After you have finished the construction of your new wilderness shelter you will need to construct a small simple "smoke house" so you can smoke all your extra wild meat and convert it into meat jerky. Later you can convert the meat jerky into pemmican. The best time to smoke meat is at night because it is more difficult for most wild animals and humans to follow a smoke trail back to your shelter at night.

You will need to harvest the summer wild berry crop and "sun dry" most of them for winter consumption. However, blackberries and raspberries do not dry well so you should learn how to make jam or wine from the wild blackberries.

Root Cellar: After you have constructed your smoke house you should consider digging a simple root cellar into the side of the hill near your wilderness shelter so you can store some of your fall garden vegetables and fruit for winter consumption. (Note: The floor of your underground root cellar should be at least 8-feet above the level of the earth at the bottom of the hill. In the event of a flood you would not lose all your food.)

Advantages of a Stainless Steel Meat Grinder

Following are some reasons why you might wish to consider investing in a **stainless steel** meat grinder. A hand-cranked meat grinder has the following advantages:

1. Converts tough cuts of meat and wild game meat into hamburger consistency.
2. All red ground meat looks approximately the same.
3. Ground meat will minimize the number of future complaints you will hear from your group.
4. Ground meat is easier to use in a wider variety of recipes.
5. Ground meat more easily and completely absorbs any seasonings you may have.
6. Ground meat cooks faster and more thoroughly than thicker cuts of meat.
7. Ground meat is easier for most people to chew and it is easier for most people to digest.

You will also need some meat seasoning tenderizer, some chili seasoning, some taco seasoning, some Sloppy Joe seasoning, and any other seasonings that your family normally enjoys.

A Few Tips on How to Grind Meat

1. Chill the meat before grinding. Cold meat is easier to grind than meat at room temperature.
2. Cut the meat into cubes between one-half inch up to one-inch in size.
   a. If the meat does not contain much fat and you are adding fat from another source then cut the meat into one-half inch cubes.
   b. If you are blending two or more different types of meat together then cut the meat into one-half inch cubes.
3. Tender meat should be coarse ground.
4. A coarse grind is preferred for hamburgers because it helps the meat hold together while cooking.
5. Tough meat or meat with lots of sinews should be fine ground.
6. A fine grind helps the meat absorb the meat seasonings and to blend more evenly with other recipe ingredients.
7. If the meat is still too tough after grinding then put the ground meat through the meat grinder a second time.

Chapter Nine

Practical Cooking Options in a Wilderness Environment

You may prepare your meals over a small cook fire, or with a solar oven, or with a microwave oven. You may use your current cook pots but you will probably want to invest in some cast iron cookware for a variety of good reasons. You could eat off of your current dishes but you may want to purchase some stainless steel dishes or some enamel coated camp dishes for a variety of good reasons. You will be eating off your current kitchen table or off of a high quality heavy-duty folding table. You will sit in a very comfortable folding chair or any chair you choose to bring with you. In other words, when it is time to eat you will be enjoying your meal in a fashion that is very similar to the way you currently eat a meal in the civilized world. The one major difference is that you will be eating in the middle of a hostile wilderness environment.

Cooking Food versus Heating Canned Food

There is a big difference between cooking food and simply heating canned food. When you first arrive in the wilderness you should have a reasonable supply of canned food that you can simply heat and eat. However, you will also be adding fresh fish and wild game to your diet and those items will need to be cooked and not just heated. And if you wish to enjoy fresh bread or a hot pizza, then those items will also need to be cooked. Therefore you will need to be able to heat canned foods and you will need to be able to cook raw food.

Canned food could be heated inside a solar oven if the sun is shining. If you have solar power then could heat your meal briefly inside a microwave oven. If neither of these options is available then you could build a very small fire inside a small hole in the ground (outside your shelter) and heat your food over the coals of that fire.

On the other hand, if you have fresh wild game meat then you will need to thoroughly cook that meat to avoid serious health problems. This is where an instant-read meat thermometer will help you determine if the center of your meat has reached a temperature that will kill all the potentially hazardous microorganisms that might be present in the raw meat.

There are several options for cooking food and for heating canned food:

1. a solar oven,
2. a microwave oven,
3. an electric single-burner hotplate (or an electric skillet),
4. an electric toaster oven,
5. a propane camping two-burner stove or a traditional propane range/oven,
6. an outdoor propane gas grill,
7. an outdoor charcoal grill,
8. a cast iron wood burning fireplace/heater with a two-burner stove,
9. a wood burning old-fashioned indoor brick fireplace/oven combination,
10. a traditional outdoor wood campfire.

Solar Oven: A solar oven is an excellent option whenever the sun is shinning. The solar oven may be used outdoors, or it may be used indoors when placed in direct sunlight in front of a south facing window. To construct a solar oven you will need 75 square feet of Reynolds Heavy-Duty Aluminum Foil (18 inches wide and 16.67 feet long) and one box of 5 Reynolds Oven Bags. Two web sites that provide simple easy-to-follow instructions for building your own solar oven are:

http://www.cookwiththesun.com/wood-box.html

http://www.oksolar.com/samples/sunoven1.html

However, a solar oven will not work when it is snowing, raining, if the sky is overcast, at night, or in the early morning or late afternoon. Therefore you will not be able to rely exclusively on a solar oven. But it would be a very good option when the sun is shining.

Solar Power: In my opinion you will need some type of solar power in the wilderness. If you have one or more solar panels, and one or more 12-volt deep-cycle batteries, and an inverter then you could use a microwave oven to quickly heat canned foods. However, you would need a huge solar system to use an electric hot-plate, or an electric skillet, or an electric toaster oven. Although these options are possible in the wilderness they simply are not practical and therefore they should not be seriously considered. (Note: This includes the small 12-volt skillets sold at truck stops.)

Propane (LP Gas) Stove: Do not buy an outdoor propane gas grill because the smell of your food will drift for miles on the wind and this is generally not desirable. A smaller propane two-burner cook stove could be used to heat food and to completely and thoroughly cook food, but it cannot be used indoors. However, a traditional propane cook stove can be used indoors. There are two basic options: a $382 Maytag brand 22-inch wide three-burner stove/oven like the ones used in R.V. campers (picture lower right), or a $550 traditional free-standing propane stove/oven with pilot lights similar to a natural gas oven currently used in many homes (no electric clocks, timers, or electronic igniters). Either of these two options will require a large propane tank with a pressure regulator. The propane tank must be located outside your shelter and a propane line run into your shelter.

Propane Camp Stove

Propane 3-Burner Stove/Oven

Charcoal Grill: A charcoal grill is another option for cooking in the wilderness. Place a layer of sand or dirt in the bottom of your charcoal grill to help protect the metal surface from the heat and to extend its useful life. Instead of burning charcoal in your grill you could burn wood until the wood forms a bed of red hot coals and then cook over those red hot coals. One potential disadvantage of a charcoal grill is that the smell of your cooking food could travel for miles on the wind and that delicious smell will lead wild animals and people to your cooking location. However, if you cook inside a covered cast iron skillet or a cast iron Dutch oven then most of your cooking aromas will remain trapped inside the cook pot and this is highly desirable when living in the wilderness.

Cast Iron Stove/Heater

Cast Iron Wood Burning Stove/Heater: A wood burning "Franklin Fireplace" or a 26-inch high "two-burner stove/heater combination" would allow you to heat your shelter and cook at the same time. If you select an appropriate wilderness area then collecting firewood should not be a problem except for the labor involved in harvesting it. However, you would need a stovepipe that extends through the roof of your shelter and it would need to be properly and correctly insulated to prevent catching your roof and your shelter on fire. The disadvantage of this option is the column of chimney smoke that would lead everyone directly to your new shelter. Finally, during the warm weather months you would need to temporarily disconnect the stove from its stovepipe and move the stove outside and put it below a simple shed type roof structure to keep it dry when it rains. This is what your ancestors did back in the 1800s in order to keep their home from becoming unbearably hot during the warm months when they used their stove to cook their meals. When the weather turns cold in the fall then you will need to move the stove back indoors and reconnect it to your original stovepipe as before.

Wood Burning Brick Fireplace: If you are a brick mason then you could build an old-fashioned combination brick fireplace/oven with a chimney through your roof. This has the same advantages and disadvantages as a cast iron stove. However, you cannot move a brick fireplace outdoors during the warm weather months.

Cooking with a Traditional Outdoor Campfire (follow all campfire safety rules to avoid forest fires): This option will work for almost everyone, and it doesn't cost anything, and it does not require any special equipment. However, I needed to first discuss all the foregoing alternatives because one of them may be a better choice for someone who is in a very unique situation. The preceding information would also enable a person who was considering one of those options to better understand its advantages and its shortcomings.

But for the vast majority of people an outdoor campfire will probably be the best choice for cooking. If you use dry seasoned hardwoods to build your cook fires then they will produce good heat and be almost completely smokeless. (Note: This is also the type of firewood you should use if you have an indoor fireplace.)

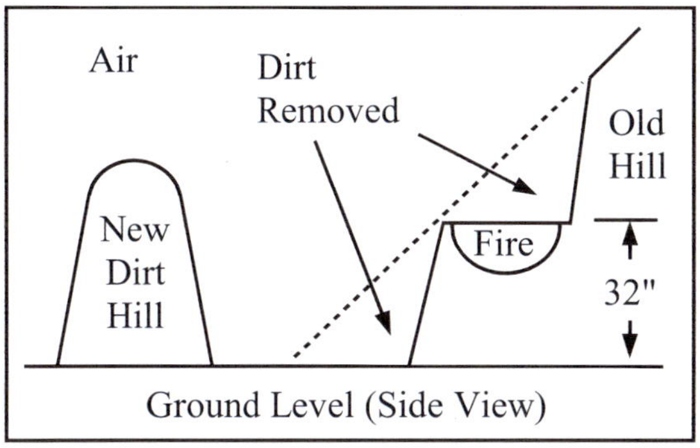

The best location for your campfire would be a cavity that you dig into the side of a hill that is at least 10 feet away from your shelter. The cavity should be about four-feet wide and four-feet deep into the side of the hill. The campfire itself should be built inside a hollowed out cavity inside a ledge about 32 inches above the ground to make cooking easier. The sides of the hill will provide a natural windbreak on three sides of your campfire and it will allow the firewood to burn down into a bed of red hot coals for cooking purposes. The fire will not be visible from any direction except directly in front of the fire. If you transfer the dirt you dig out your new fireplace opening to a position about four feet in front of the campfire opening and build a natural dirt wall about four feet high then your campfire would not be visible from this direction either. You should also consider building some type of rainproof covering about seven feet above your campfire to keep the rain off you and your campfire. Do not completely enclose your campfire because you do not want to create an oven.

You will need to learn how to cook over the coals of a campfire. A covered cast iron skillet or a covered Dutch Oven will keep the heat inside the cookware to more evenly and quickly cook your meals. You can also bake bread, pastries, and pizzas inside a covered cast iron skillet or Dutch Oven. They will keep most of the smell of your delicious food from escaping into the wind to lead wild animals and people to your wilderness shelter.

Folding Camp Oven: Another option is a camp cook stove oven. The picture of the oven is a folding metal oven and it can be unfolded and supported on four flat rocks above the red hot coals of a campfire (not directly on the coals). You can bake almost anything inside this oven. The bigger camp ovens have the same appearance but they do not fold up and they are useful for baking larger items. The disadvantage of these ovens is they release the delicious smell of your baking food into the wind.

Cast Iron Cookware

Cast iron cookware is the recommended cookware for use on a cast iron stove or inside a wood burning fireplace or over a wood burning campfire. Cast iron cookware is 100% cast iron and it does not contain any parts that are made of plastic or aluminum or wood that could melt or catch fire when exposed to the intense heat of a wood burning fire.

I suggest you purchase cast iron cookware with a lid for two reasons:

1. The lid will keep embers and ashes and soot from getting into your food inside the cook pot.
2. The lid will keep more of the heat and the flavor of the food inside the cook pot and this will minimize the amount of firewood needed, reduce the total time required to cook the food, and maximize the flavor of the finished food.

Six Quart Dutch Oven with Lid

Some cast iron cookware, such as a Dutch Oven, is sold with the lid as part of the unit. However, most of the other cast iron pieces are not sold with a lid. Instead a lid must be purchased separately. I strongly recommend that you purchase a lid for each piece of cast iron cookware that you acquire.

For additional information about cast iron cookware please read the cast iron cookware chapter in my book **"Grandpappy's Survival Manual for Hard Times"** in order to learn how to cook with cast iron and how to take proper care of your cast iron cookware.

Stainless Steel Cookware and Teflon Coated Cookware

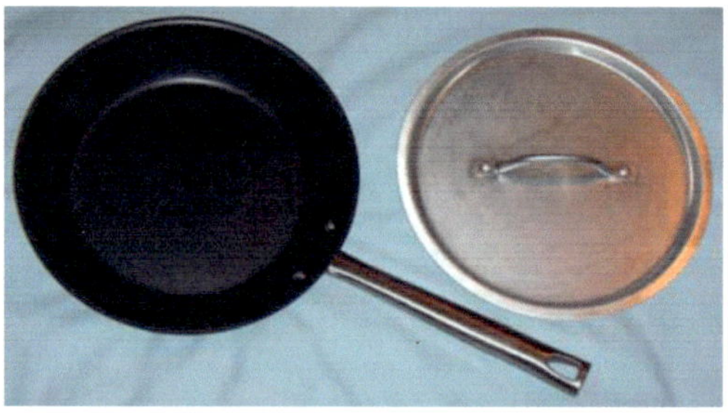

Teflon Coated Stainless Steel Skillet

If you can find 100% stainless steel cookware without any plastic parts then it could be used to cook over a wood burning fire.

A stainless steel skillet cover may be used to cover your pots and skillets and it can be safely used with a wood burning fire.

Some stainless steel cookware if very cheaply made and it will not last very long in any type of cooking application.

The advantage of cast iron cookware over stainless steel cookware is that food will usually not stick to the inside of the cast iron cookware.

Cookware that has a Teflon non-stick coating on its interior cooking surface can be used with a wood burning fire if the rest of the cookware is made of heavy-duty metal and it does not contain any aluminum or have any plastic or wood handles. This type of cookware can sometimes be found in the camping section of a sporting goods store, or a hardware store, or an Army-Navy store, or a Walmart.

Dishes and Tableware

Enamel Dish Set

You should take your current stainless steel tableware (knives, forks, and spoons) with you into the wilderness. If you have a set of carving knives then you should take them with you also.

You may also take your current dishes but you may wish to consider stainless steel camp dishes or enamel coated camp dishes. Stainless steel and enamel dishes are usually much easier to clean than regular dishes using cold water or warm water. However, if you allow sliced fruits or any high acidic food to remain for an extended period of time on your enamel dishes then that food will permanently stain those dishes in the exact spot where the food made long-term contact with the enamel. Therefore you should wipe or rinse the food residue off your enamel dishes immediately after eating a meal in order to preserve the appearance of your enamel dishes.

Chapter Ten

Propane as a Cooking and Heating Fuel

All propane tanks must be located outside your shelter and a propane line run into your shelter.

Propane Pounds and Gallons: According to the National Propane Gas Association one gallon of propane weighs approximately 4.24 pounds. This is at 60ºF (15.5ºC). Propane expands approximately 1.5% per 10ºF (5.6ºC). Therefore one gallon of propane weighs about 4.1 pounds at 77ºF (25ºC). Propane dealers may legally sell propane by either weight (pounds) or volume (gallons). If you know the above conversion factors then you can compare the prices among different dealers and different size propane tanks. The following table may help:

Weight	Volume	Weight	Volume
16.4 ounces	0.24 Gallons	100 pounds	23.58 Gallons
20 pounds	4.72 Gallons	424 pounds	100 Gallons
30 pounds	7.08 Gallons	1,060 pounds	250 Gallons
40 pounds	9.43 Gallons	2,120 pounds	500 Gallons

The small 16.4-ounce canisters of propane are a very poor value at $2.97 each because this equals about $12.38 per gallon of propane. A better value would be the 20-pound, 30-pound, 40-pound, or 100-pound tanks. You can find new empty propane tanks at any store that sells gas grills, and at some stores that refill empty propane tanks, and at places like Lowe's and Home Depot. I suggest you buy new propane tanks and then take them to a propane filling station and have them filled while you wait. Watch your tank as it is filled and make sure you get your own new tank back. Do not let the employee talk you out of your new tank by telling you that their used pre-filled propane tanks are just as good. In some cases this may be true. But in some cases it may not be true. If you buy new tanks then you will know that you have quality tanks that should last many, many years without any problems. When your propane tanks are empty you can take them to any refilling station and have them refilled while you wait. Most R.V. campgrounds have a propane refilling tank near their front office and they will refill propane tanks for anyone because it generates revenue for their business. (Note: The 20-pound tanks, such as the "Blue Rhino" in the picture, are frequently only filled with 15-pounds of propane for "safety" reasons. This is an example of reducing the contents of the package while keeping the package price stable. If you take your tank to a propane refilling station then they will fill it with 20-pounds of propane.)

Another option would be to purchase a new bulk propane tank from a propane supplier (250 to 1,000 gallons). There are two options: above ground and below ground. The above ground tanks can be filled to about 80% of capacity. The below ground tanks can be filled to about 90% of capacity because the below ground temperature does not fluctuate the way the above ground temperature does. The below ground tanks would also be virtually invisible because only a very small refilling port is above the level of the ground.

Your propane should be reserved for emergency heating or cooking purposes and you should not consume it just because it is available. Do not use your propane to fuel a propane lantern. This is a terrible way to waste a resource as precious as propane. Instead use battery operated L.E.D. lanterns or L.E.D. flashlights for light.

Propane (LP Gas) Heat: Some propane heaters may only be used outdoors. And some propane heaters **must be vented** to the outside air. Do not purchase these types of heaters.

If you are considering a propane heater then you should purchase a "**safe indoor propane heater**" that is designed to be 99.9% fuel-efficient so it can be operated indoors without the need to vent it to the outside air. There are several different types of these "safe indoor propane heaters." Three of these options are illustrated below.

The first propane heater in the left picture is portable. It may be used with the 16.4-ounce canisters of propane and it will heat a small area for between 1.5 to 6 hours. When attached to an outdoor 20-pound propane tank using a Flexible Hose and a Fuel Filter it will provide heat for between 25 to 110 hours. (Performance information is printed on the outside of the box). For safety reasons this type of heater may not be used at night when you are asleep. I have never personally owned one of these heaters so I can make no comment about its performance.

The second propane heater in the above middle picture is a Blue Flame Heater.

The third propane heater on the above right is an Infrared Heater.

Both of these heaters have a non-electric thermostat that allows you to control the temperature inside your shelter at any level you wish. Neither heater uses any electricity or fans and the heat from both heaters simply radiates out into your shelter in a manner similar to a wood burning fireplace.

Both the Blue Flame and Infrared heaters must be attached to a minimum 100-pound outdoor propane tank or larger. When properly installed these heaters may be used safely indoors 24-hours per day without any health problems.

After being shut down all winter both the Blue Flame and the Infrared units need to have their propane pilot light port cleaned of dust using an air blast from a can of compressed air (the same type of compressed air you would use to clean the keys of a computer keyboard). This is relatively easy to do and it only needs to be done once per year at the beginning of the cold weather season. After you light the pilot light the pilot light will remain lit all winter without any human intervention.

During the course of my life I have personally owned one Blue Flame Heater and two Infrared Heaters. In each case these heaters were my only source of heat during the winter months. I never had any problems with the Blue Flame Heater but I did have a very minor repair problem with one of the Infrared Heaters.

You will also need a Pressure Regulator (picture on right with red shut off value below) to control the propane pressure from your outdoor propane tank to your propane heater. These regulators are sold based on the size of the propane tank and the size of all of your propane applications.

Final Propane Consideration: Unless you have a 500 gallon or larger propane tank, or a reliable method for refilling the smaller tanks, then you should consider some other heating and cooking option.

Kerosene Heaters: Based on the negative health feedback issues I have personally heard I cannot recommend any type of kerosene heater. Therefore I will not discuss kerosene heaters in this book.

Chapter Eleven

Electrical Power Options for a Wilderness Shelter

Water Power

If you are near a waterfall that has a column of water that free falls at least twenty-feet then a water turbine would be a reasonable investment. These turbines are placed at the foot of the waterfall and a large diameter pipe is placed inside the waterfall so the water flows down through the pipe and out the turbine to create electrical power. In my opinion this is a very reasonable method for producing power, it is relatively affordable, it is extremely quiet, and it will work 24 hours per day 365 days per year if you have good rainfall in your area and the waterfall does not dry up.

Wind Power

If you relocate to a very windy area that has good strong winds all year then a wind generator would also be a feasible option. But windy climates are hard on garden crops. The disadvantage of a windmill is that it must be placed relatively high off the ground (usually at least twenty-five feet) and this makes it extremely visible for a long ways in almost every direction which may be a disadvantage depending on your specific situation.

Gasoline Generator

Do not buy a gasoline generator. They make a lot of noise even when they are muffled. And the very efficient ones will only run about 2.5 hours on one-gallon of gas. Renewable solar power is the best option for producing electricity.

Solar Power

The sun is a source of renewable energy. If you had a solar panel and a 12-volt deep-cycle marine battery or a golf cart battery, then you could capture and store the sun's energy inside your 12-volt battery. You could then tap that battery whenever you needed power. For example, you could recharge your portable computer, or your cell phone, or your AA batteries that you use in your radio or flashlight. If you also had a 1,200 watt inverter then you could run 110-volt electrical appliances for short periods of time, such as an electric drill, or a printer, or a small microwave oven. This means you could heat some canned food to a reasonable temperature inside a small microwave oven using the sun's energy you previously stored inside a 12-volt deep-cycle battery. Solar power is such a practical option that every family should have their own solar generator.

A solar generator can be easily constructed by almost anyone using the following items for a total cost of about $275:

1. One 15-watt monocrystalline solar panel. These panels are sometimes sold with an optional charge controller attached to the wiring to prevent overcharging of the 12-volt battery. ($75)
2. One 12-volt deep cycle marine battery or golf-cart battery. ($80) (Note: During a serious emergency you could use your automobile battery but your car battery is not a deep-cycle battery so it would not store as much energy as a marine battery or a golf cart battery.)

3. One 12-volt cigarette lighter adapter with three outlets for a cell phone, or portable computer, or similar devices. ($17)

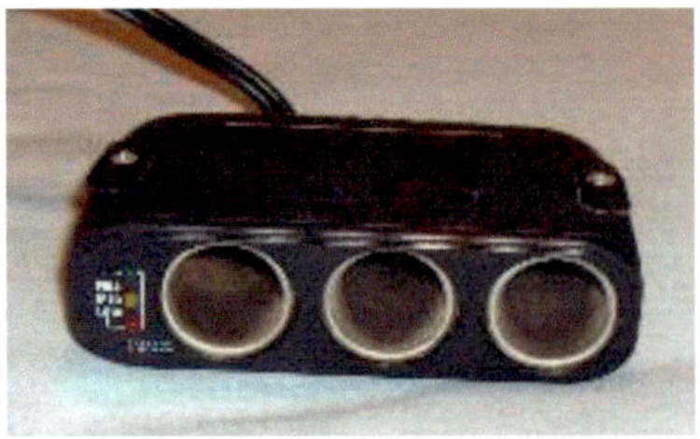

4. One 110-volt inverter that provides at least 1,200 watts of continuous power or 2,400 watts of peak power. An inverter normally has either one or two of the three-prong 110-volt outlet sockets built into the front panel of the inverter so you can easily plug your 110-volt appliances directly into the inverter without making any modifications to your appliance plugs. ($100)
5. About ten feet of 12-gauge electrical wire in two different colors such as red and black. ($3)

Connect the two alligator clips from the solar panel to the positive red wire and negative black wire terminals of the battery. Connect the two color-coded wires from the 12-volt adapter to the battery. Connect the 110-volt inverter to the battery using some of the 12-gauge electrical wire.

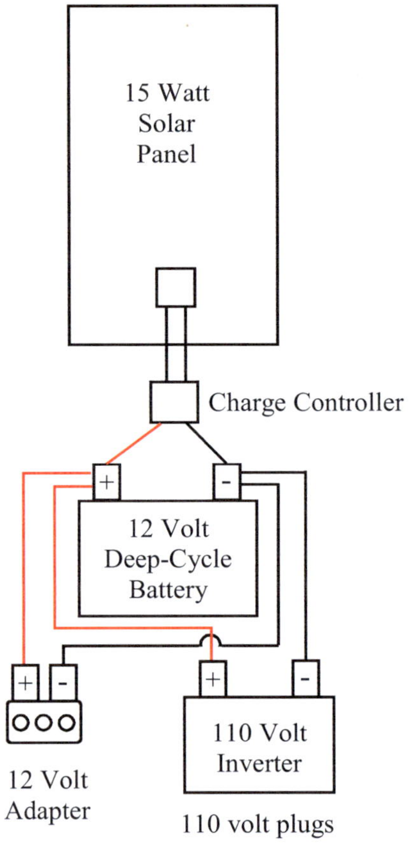

Entertainment: You should have a portable computer (laptop or notebook) with a DVD player and your entire collection of DVDs. A portable computer is a very low power consumption device. With these basic items you would be able to watch one of your favorite movies every evening for entertainment. You should also have your favorite music files saved on the hard drive of your portable computer (and backed up on a DVD disk). And you will still be able to play your favorite computer games. The ability to play your favorite games and watch your favorite movies will help to minimize the impact of "cabin fever" and it will give you a better chance of surviving the mental strain of wilderness living. If you have a spare portable computer with a DVD drive then you should also take it with you as a backup. A portable computer is not very big but it can provide thousands of hours of entertainment possibilities.

Current News Events: You should have a battery operated world-band radio. You will be able to listen to the music of your choice. And you will be able to stay informed about the local, regional, and world news events. You will also be able to listen to a wide variety of talk-radio shows from around the world. This will help to keep you from feeling isolated and uniformed. You should also have a simple battery operated AM-FM radio. With two (or three) radios you will have the ability to stay informed about world events if something should happen to one of your radios. Just knowing what is happening in the rest of the world helps to minimize the feeling of isolation and loneliness.

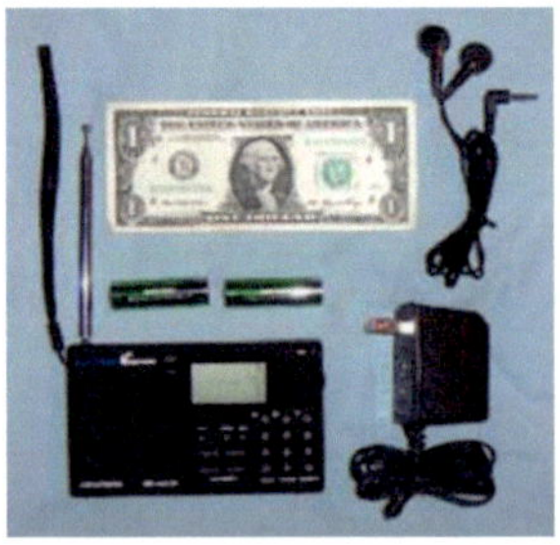
Grundig G6 Radio

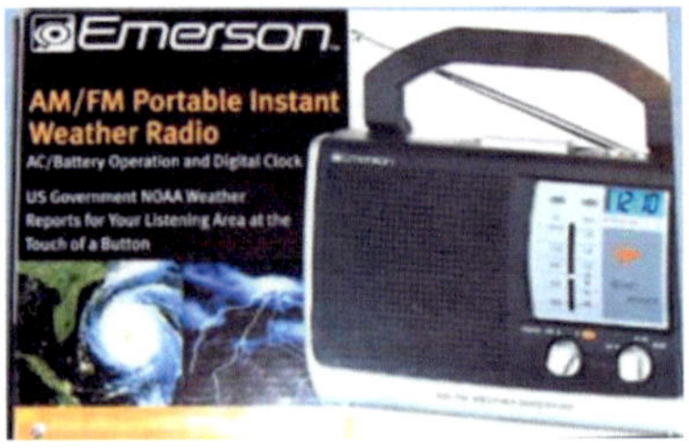

Communication: Depending on your wilderness location, you may or may not have a cell phone. A cell phone would keep you in touch with the outside world but you would have a monthly expense with most cell phones (a Tracfone can be prepaid annually). It is also possible that a cell phone tower may not be in range of your new wilderness location. Another more reasonable alternative would be to purchase a "satellite phone" so you can receive telephone calls that are relayed through a space satellite. However, this convenience would involve a regular monthly expense so you would need to make this decision based on your resources and your needs. An amateur ham radio is another possibility but it does require a license. If you truly wish to disappear into the wilderness then you should also remember that your position could easily be triangulated if you use any type of mobile phone or ham radio.

Light: You should have an L.E.D. high-intensity flashlight so you can see deep into the darkness all around your camp if it becomes necessary. You should have a reasonable supply of rechargeable batteries that you can recharge using a solar battery charger. You should have a solar-power L.E.D. camp lantern so you can have light at the flip of a switch. The L.E.D. lantern could be hung in the center of your living area until bedtime. Then the lantern could be placed at the head of your bed when you retire at night. Being able to easily turn on the light in the middle of the night helps to minimize the psychological strain of wilderness living.

Solar Battery Charger: You should also have a portable solar battery charger (see bottom of this page) that will charge the most common types of NiMH batteries, such as AAA, AA, C, and D. You will also need a reasonable supply of rechargeable batteries for all your batteries operated electronic items, such as your flashlights, lanterns, radios, and electric toothbrushes.

Rechargeable NiMH Batteries

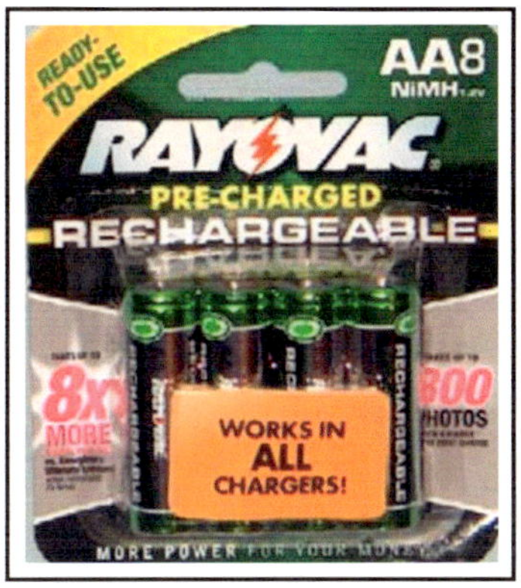

When possible you should allow your rechargeable batteries to fully discharge inside the device in which they are being used before you recharge them. In other words, wait until the electrical device either stops working or it tells you it is time to replace the batteries. This will mean the batteries are fully discharged and they are now ready to be recharged.

It should also be mentioned that a NiMH battery does not have the "memory" problems of a Ni-Cad (Nickel-Cadmium) battery. The original Ni-Cad batteries had to be charged to their peak potential the very first time they were charged or they would remember the first charge level and you would never be able to charge that specific battery in the future above that original charge level. NiMH batteries do not work this way and if you don't fully charge a NiMH battery then it only impacts that one usage and it has no impact on the future maximum charging potential of that battery.

Most of the time you will probably charge your rechargeable batteries to their maximum power level. However, it is nice to know that if an unexpected emergency occurs and you do not have the time to charge your batteries to a full charge, then it is okay to use those partially charged batteries and it will not impact the future charging potential of those batteries. This means if you desperately needed your batteries after the sun has gone down or if it is raining, then you could use your batteries in your flashlight or radio even though they were not yet fully charged and you would not be damaging those batteries in any way.

Solar Battery Charger for Two Batteries

This two battery solar charger (picture on top of next page) was originally called the "Solar Eleven-in-One Battery Charger with Meter." However it is currently called a "Solar Battery Charger." Although the name has been changed these two chargers are exactly the same. This unit will charge the seven "GUM" size battery that are used in Europe. But the box label does not mention this because it is not important in the United States.

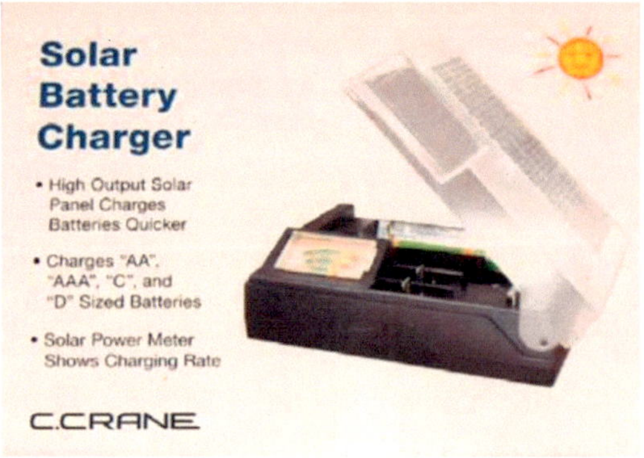

This Solar Battery Charger has a meter that shows when you have the solar panel at the optimum angle for recharging your batteries. The meter eliminates the guess work of whether or not you have the solar panel at a good angle. Just keep changing the angle of the solar panel until the meter needle reaches its highest point and then starts going in the opposite direction. The meter will not work until after you have inserted two batteries in the unit. The meter is connected to a circuit from the solar panel through the battery compartment and until you have two batteries inserted inside the unit that circuit is not complete. (Note: If you wish you can check the meter in about two hours when the position of the sun has changed and reposition the solar panel to continue to use the sun in the most efficient manner. However, the solar charger will still continue to charge your batteries even if you don't adjust the angle of the solar panel.)

Solar Battery Charger for Four Batteries and for Some Cell Phones

This solar battery charger will recharge four batteries at the same time. It comes with a special cord that plugs into the charger and the other end of the cord has five standard plugs for recharging the batteries in some cell phones.

There are at least ten different size plug connectors that are used on cell phones and other rechargeable electronic devices, such as portable computers. Consequently there is no guarantee that this unit will be compatible with your cell phone or other rechargeable electronic equipment.

The solar panel on this four-battery charger is approximately 66% bigger than the solar panel on the two-battery charger. Therefore it can charge four batteries in almost the same amount of time as the two battery charger.

Suggestions on How to Use a Solar Battery Charger

Batteries are not designed to be left in direct sunlight for an extended period of time. They will overheat and start to leak and you will get some very nasty stuff on the inside of your solar charger. Batteries react to the sun the same way your skin does. If your skin is exposed to direct sunlight for too long then you will get a very painful sunburn (or worse). To avoid the sunburn you could sit in the shade. Therefore the simple solution is to put a white cloth over the batteries, between the batteries and the solar panel, which is on a hinge. Then elevate one end of the white cloth with a short stick or piece of plastic to allow the batteries to breathe.

The chargers will work on either two or four identical batteries of exactly the same brand name, of the same design, and of the same size, at the same time. Do not mix batteries of different brands at the same time. And it is not a good practice to try to charge batteries that are not equally discharged. Therefore I normally allow my batteries to get very close to a full discharge before I stop using them. Then I can safely place them in the solar charger and each battery will receive an equal charge.

It would probably be wise to invest in several rechargeable batteries of the same size. This would allow you to always have spare batteries while your discharged batteries are in the solar charger. And there will probably come a day when some emergency situation demands your complete attention for an extended period of time and you will cook a set of batteries. Therefore it is nice to have spare batteries.

It would also be nice to have a spare solar battery charger in the event some battery chemicals get all over the inside of your charger and it is damaged beyond repair.

Chapter Twelve

Shelter Alternatives for Long-Term Wilderness Survival

When people first consider living in the wilderness their first thoughts are usually about living inside a tent, or a van, or a recreational vehicle (camper). Therefore let's begin by examining all three of these obvious shelter options.

Tents

Tents are designed for use as a temporary shelter for a short period of time, and for easy transportation between locations, and for use in three of the four seasons. Therefore you will not need a tent. This includes the expensive canvas tents that are advertised for year round use and which have a "stove pipe port" in the roof of the tent. Tents provide almost no protection from severe winter weather and they are not energy efficient when it comes to heat conservation. And they provide no protection from aggressive wild animals such as bears. If a bear smells food inside your tent then the bear will collapse the tent on top of you and claw its way into your shelter. You will be trapped inside a huge piece of collapsed canvas and you will be ripped to pieces as the 500-pound bear searches for its next meal. This is not a scenario you want to expose yourself to. Therefore my recommendation is to not invest any money in a tent. However, that is just my advice and you may do whatever you wish.

Vans

A van would be a reasonable temporary living shelter during the spring, summer, and fall because you can put a full or queen-size mattress in the rear of the van and get some good sleep every night. However, since you cannot stand straight up inside a van it would be extremely uncomfortable when you are trapped inside your van on rainy days. Finally, a van would be totally unacceptable during the freezing winter months because you would freeze to death inside your van.

A Recreational Vehicle (R.V.) or a Camper

If you already own an R.V. or a camper then it would be the ideal temporary home during two or three of the four seasons. It would obviously be acceptable during the spring and autumn. And if it is not parked in the direct sun and if the summer temperatures do not get into the 90s or the 100s then it would also be fine during the summer. However, if the summer temperatures are really high then the inside of your camper will become like an oven and you will be miserable.

The same concept applies to the winter months. During freezing cold weather the inside of your camper will become like a food freezer. These temperature extremes are not acceptable for human habitation.

Therefore one of the most serious problems with a camper is that the inside camper temperature gradually adjusts to the outside air temperature due to the lack of superior quality insulation. This temperature adjustment process normally takes about two-hours but it will happen. If your objective is to survive and thrive in the wilderness then a camper would not be the best choice for year-round comfort.

Another major disadvantage of any type of camper is that it does not contain enough empty storage space to stock several years worth of supplies inside the camper. Therefore you would still need to construct some type of nearby small underground root cellar to store your long-term food supplies in a safe temperature controlled environment.

The final major disadvantage of a camper is that you may not be able to get it into the wilderness area that you select if that wilderness area is not on some type of suitable access road.

However, a camper may have practical value as a year-round home if you are able to park it beside a very steep hill and then gradually cover the camper with dirt from the hill. I am fully aware that this idea sounds ridiculous but it may actually work. There are several major issues that would have to be resolved such as: (1) providing an air-intake line to your propane heater, (2) moving your propane tanks to a new location near the front door of your camper and modifying the propane line accordingly, (3) moving your 12-volt batteries to a position below and behind your front steps and running the electrical cables accordingly, (4) providing a way to safely open and close your gray and black water tank values to periodically empty those waste water tanks, (5) providing a garden hose attachment from the front of your camper to your fresh water tank that is downhill (gravity) the entire distance from the front of the camper to the water tank, and (6) protecting the exterior of your camper with high-quality heavy-duty tarps before your cover it with dirt. You will also need to protect your roof antennas from damage by protecting them or removing them. Obviously you would need to leave an entrance to your camper door and to any windows on the door side of your camper.

Another option would be to dig a rectangular hole into the side of the hill using a rented backhoe. The hole should be about four feet wider than your camper and about four feet longer than your camper. Then you could build a wooden frame structure inside the hole that would completely surround your entire camper. The wooden frame structure should be about two-feet taller than your camper to provide a crawl space so you can do any necessary maintenance to the roof of your camper if the need should arise. You should nail plywood on the outside and tops of your wooden frame structure. Then you could use the backhoe to transfer the dirt you removed from the hole on top of the frame and on the outside of the frame. Then you could back your camper into the middle of your new wooden frame and you would have about two-feet of clearance all the way around the outside of your camper to do routine maintenance on your camper. But this involves almost as much work as simply building a new underground home. One serious disadvantage is that it would be really dark inside your camper. The primary advantage of this strategy is that you would quickly have a fully operational comfortable underground shelter that is hidden from view and that would stay at an acceptable temperature all year. You will have also become invisible in a reasonably short period of time.

Shelter Considerations

When you think about your future shelter you should think in terms of long-term comfort. Your shelter will need to be tall enough so you can easily stand up inside your shelter and move around. Your shelter will also need to be wide enough so you can comfortably lie down in order to get some sleep. However, your shelter will also need to be relatively compact in order to conserve heat. The larger your shelter the more difficult it will be to keep it at a comfortable temperature during the bitter cold winter months. Therefore you will need to reach a strategic compromise between being too big or too small.

A shelter with 7-feet (84-inches) of clearance from floor to ceiling instead of the traditional 8-feet (96-inches) would allow enough clearance for a standard size doorway, and it would reduce the volume of air inside the shelter by 12.5%. Look around the room you now live in. What do you see within one-foot of the ceiling all around your room? Probably nothing. Just wasted space that you pay to heat and cool. If you eliminate this wasted space in your wilderness shelter then your shelter will be a little warmer in the winter (heat rises).

Finally your shelter should provide you with a feeling of safety and peace of mind when you are inside your shelter. You should not have to worry about being unexpected attacked by a wild animal. You should not have to worry about being seen by other people unless you choose to attract their attention to your shelter. Finally, you should be able to see outside your shelter so you can determine the current weather conditions, and you should also be able to see what made that noise you just heard. Either a "wilderness cabin cave" or an underground home will satisfy all the above requirements.

Cabin Cave

A cabin cave or an underground home that is built into the side of a steep hill would be far more practical than an above ground home or a traditional log cabin or any type of tent or camper. During the winter an underground shelter that is built into the side of a steep hill will be warmer and during the summer it will be cooler. While you are building your permanent shelter you could sleep inside your vehicle or inside a safe temporary shelter. Complete instructions for building a safe temporary shelter in about two or three days is in my book **"Grandpappy's Survival Manual for Hard Times."**

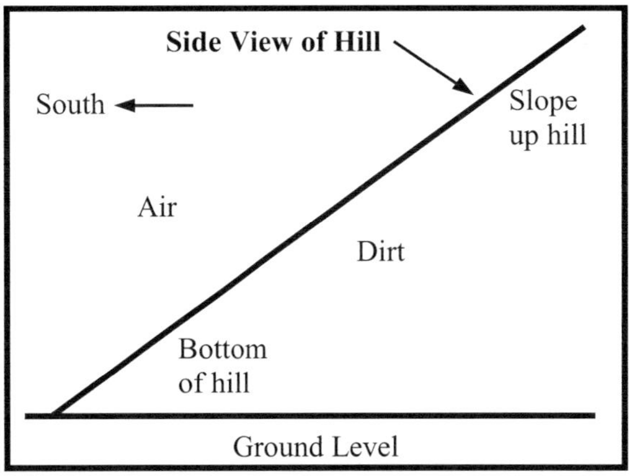

A cabin cave should be built into the side of a hill that has a slope of about 45 degrees. You would need to dig your cabin cave hole straight down into the side of the hill and not sideways into the hill like a mine shaft. Complete detailed instructions for building a cabin cave are also contained in the above book.

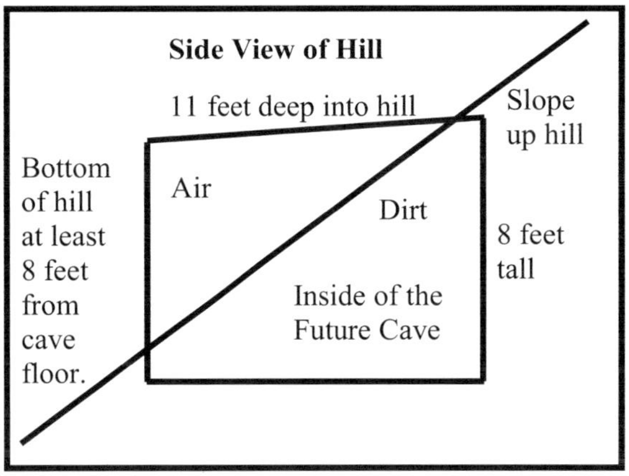

Following are some of the advantages of a cabin cave when compared to a camping tent, a camping trailer, or a traditional log cabin (the following list was copied from the above book):

1. The exterior of a manmade cabin cave would be of an irregular shape and it would blend in with the surrounding environment. Therefore, unlike a tent, trailer, or cabin that could be easily seen from a distance, a manmade cave would be almost invisible. This would provide its inhabitants with a significant degree of safety. They would also be virtually invisible to modern heat seeking technology.
2. The location of most natural caves and manmade mine shafts are well known to the residents living in the surrounding area and to the government. However, you would be the only one who would know where your cabin cave was located.
3. A carefully hidden cabin cave would be a relatively safe place to store your equipment, supplies, and food.
4. You could snore, sneeze, talk, laugh, or make a little noise and no one outside the cave would hear you.

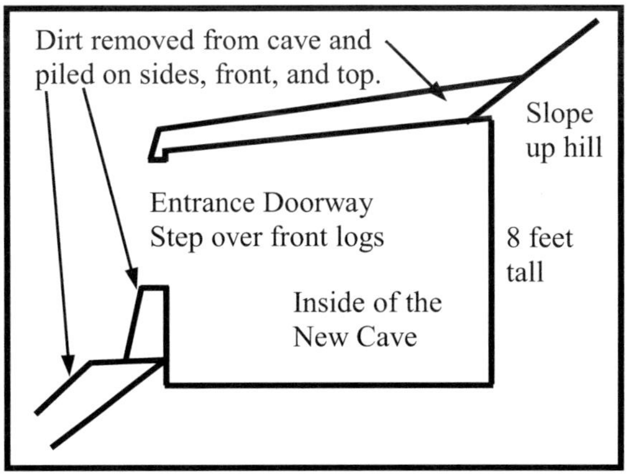

5. You would not have to worry about a rifle bullet penetrating the walls of your cabin cave.
6. The inside of your cabin cave would remain at approximately 50 to 60 degrees Fahrenheit all year (10 to 16 degrees Celsius). This is a comfortable temperature in the winter, spring, summer, and fall. You wouldn't need to keep a fire burning during freezing winter weather or on very cold nights. You would only need to hang a thick blanket over the inside and outside of your doorway entrance. This would help to block air drafts and it would keep the natural warmth inside your cabin cave. And during the summer months you would be able to stay cool without electrical fans or air conditioners. This means you could get some well-deserved sleep every night regardless of the time of year. (Note: The 55 degree Fahrenheit (13ºC) average year-round temperature is the reason bears, groundhogs, and many other animals hibernate inside a cave or a deep hole in the ground. It keeps them from freezing to death during the winter.)
7. You would not have to prepare a huge stack of firewood in order to comfortably survive the winter. You would only need enough wood for a small daily cook fire. And perhaps an occasional small fire after returning to your cabin cave if you went outside for an extended hike during freezing winter weather. If these fires were built from well-seasoned hardwood then these fires could be small and virtually smokeless.
8. Since you would not need to keep a continuous fire burning all winter then there would be no column of chimney smoke to attract attention to the location of your cabin cave.
9 The 55 degree Fahrenheit (13ºC) average year-round inside cave temperature would help to extend the shelf life of any food you may already have and it would extend the shelf life of any future wild game meat you convert into smoked meat jerky or pemmican. This temperature would also be appropriate for the winter storage of root vegetables such as potatoes and for whole fruits such as apples and pears.
10. A cabin cave would provide excellent protection from all types of wild animals. A wild animal may be able to dig through the dirt but it could not get through the 4 to 7 inch diameter logs. Any animal trying to attack you by breaking through your doorway entrance log door could be killed by sticking a sharp spear deep into its face, neck, or body.
11. Since a cabin cave is below ground it would be a relatively safe place to be during a tornado, hurricane, or some other type of really bad weather.
12. Finally, a cabin cave may also provide some reasonable degree of protection from nuclear fallout in the event of World War Three.

Underground Home

Another option would be to purchase and transfer some plywood, some 2x4 lumber, and a few cinder blocks to your wilderness location. You could use the cinder blocks as a foundation, and the lumber, plywood, some roofing paper, some roof shingles or roofing metal, some nails, and some heavy-duty tarps to quickly build an underground shelter in a manner similar to a "cabin cave" except you would be using lumber instead of logs. This would save you a tremendous amount of time and work because you would not have to cut down pine trees and haul them to your building site. You would only need to dig out an area of the proper size in the side of the hill where you plan to build your new home. If you adopt this idea then you should also purchase two good large glass windows (with winter storm windows and summer screens), and a full size metal front door for your new underground home that you will build into the side of a hill. Install the door in the center of the south facing wall between the two windows. You could then open the windows in the warm weather to adjust the inside temperature of your home, and you could close them in the winter to keep the natural heat indoors. The windows should be relatively big and of the type that you can raise and lower, and which also have an exterior storm window that can be raised and lowered. The reason for the large size is because you will only have two windows in your home and they will be your only source of sunlight during the day, and moonlight after dark.

You would still need to pack the dirt around the other three walls and onto your roof in order to gain all the advantages of a "cabin cave" as described earlier in this chapter. Your heavy-duty tarps should be installed between your home and the dirt. If you want to hide your home then you should also pile some dirt against a heavy-duty tarp along the front south side of your home but leave the door and the two windows uncovered. You could then transplant several evergreen bushes in front of the south wall to help conceal the door and the two windows.

Additional Shelter Considerations

The best type of shelter for both summer and winter would be either a cabin cave or an underground home built into the side of a hill. It will take some time to properly construct a cave or home of this type but once it is finished you will have a safe, secure place to sleep and to store your possessions.

Ideally you should arrive at your wilderness location in the spring or early summer. This will give you a few months to establish a permanent shelter. By the time winter arrives you should have finished construction of your below ground living shelter. Living inside a manmade cave is not as unpleasant as you might think. For generations many families built storm shelters and root cellars and they used them during extreme bad weather.

Constructing your new shelter should be your top priority project and you should devote at least 90% of your available work time to building your new shelter. Only invest a little time each day to fishing, hunting, trapping, gardening, or doing other activities. When you complete work on your shelter and just before you move into it, take the time to spray the entire inside of your shelter with commercial strength bug poison to minimize future insect infestation problems. If you don't have professional strength insect poison then any bug spray, including RAID or a generic brand, is better than nothing.

Backhoe: If money is not an issue then you could rent a backhoe and dig three holes into the side of the hill at least 8 feet above the bottom of the hill. You could dig a big cavity for your future shelter, and a smaller one for a root cellar, and a small one for a smoke house. And you could dig a fourth hole about 25 feet away from your shelter into the ground for an underground propane tank (consult your local building codes). Then you could purchase (not rent) either a 500-gallon or a 1,000-gallon underground propane tank and have it filled. In most areas you can fill an underground propane tank 90% full. You would also need to install a pressure regulator at the bottom of an outside wall on your shelter, and install copper propane lines into your shelter. Then you could install a "safe indoor propane heater" and you would have emergency heat during the worst part of the winter. You could also install a standard propane range/oven and you could cook with propane.

Transportation: Depending on how much food, furniture, equipment, and supplies you accumulate you may need to make several trips from your current location to your future wilderness home. One way you can reduce the number of round trips is to select your future wilderness building site first and then purchase some of your supplies, such as some of your food and all of your building supplies, at a town that is relatively near your future wilderness home. While your are building your shelter you could rent a storage locker in a nearby town for your furniture.

Vehicle: A SUV or truck would be fine. However, if you have a mini-van or a standard size van then you could remove the rear seats and store a lot of your supplies inside heavy-duty plastic totes and put those totes on the floor of your van. You could then put a piece of plywood on top of those containers and place your mattress on top of the plywood. You would now have a comfortable place to sleep inside your van at night. And later you will have a piece of plywood for a construction project. With this in mind, it would probably be a better idea to purchase 3 or 4 one-quarter inch thick sheets of plywood instead of one single three-quarter inch sheet of plywood. Later you would have three or four sheets of thin plywood for a future building project.

Trailer Hitch and a Flat-Bed Trailer: If your vehicle does not already have a trailer hitch then you may wish to have a trailer hitch professionally installed onto the rear of your vehicle. This will allow you to attach a flat-bed trailer to your vehicle. You could then stack and secure between 1,000 to 2,000 pounds of supplies onto this trailer, depending on the maximum load rating of the trailer you purchase. This should easily be within the legal load limit for towing a trailer with your vehicle. You could then secure one of your heavy-duty tarps securely over your supplies so your supplies will stay covered and watertight while traveling to your future destination. A covered utility trailer would also be an option, but the sides and roof of an enclosed trailer usually add another 500 to 1,000 pounds to the empty trailer weight and this would reduce the actual weight of supplies you could pull with your vehicle. The trade-off is that your supplies would be locked inside the trailer and the trailer would also provide some extra protection from the wind and the rain. If you have an SUV or a truck then a covered utility trailer would also provide a place to put your mattress and a place to sleep at night while you are building your future shelter. You should think carefully about the trailer option because you may be better off without a trailer depending on your specific circumstances. The money you would have invested in a trailer may be better invested in other items. That decision is up to you. If feasible, buy a used trailer in good shape. Or rent a trailer to move your belongings to your new home and then return the trailer to the rental store.

12-v Ceiling Light

Comfort: It would be relatively easy to expand on the above idea and convert the underground shelter into what most people would consider to be a very comfortable wilderness home. Install a few solar panels, some 12-volt deep-cycle batteries, and an inverter. Then install some electrical wiring and some switches and a few camper 12-volt ceiling lights, and you would have 12-volt electrical lighting inside your shelter just like a camper. Install a couple of 12-volt cigarette lighter outlets and you could easily recharge your computer and your cell phone. And if you invested a little money in some white or very light color wall paneling (to reflect light and keep the interior cheerful looking), then you could have a very attractive interior. Add some white paneling on the ceiling and a piece of vinyl on the floor and you would have a nice looking home. Put down a rug or two, put a few pictures on the wall, and a vase of artificial flowers on the table, and you are done. And you have just won the battle against the wilderness and you are now ready for whatever comes your way. You are no longer just surviving -- you are thriving.

Furniture: You could use some of your existing furniture inside your new wilderness home. For example, you could use a very comfortable sleeper sofa, a kitchen table and chairs, a dresser with a mirror, a chest-of-drawers, and a queen-size bed. Everything could be logically arranged inside one huge living area, or you could separate your living quarters into two rooms, a combined kitchen and living room, and a small separate bedroom. The two rooms could be divided by a wall or by a large tarp or heavy blanket that is hung from the ceiling to the floor. (Note: A floor plan sketch of this type of shelter is in Appendix One of this book.)

Sleep: Getting a good night's sleep is critical to maintaining your physical and mental health. In addition to a regular mattress you should have a really good cold weather rectangular shape sleeping bag and not a mummy bag. Two individual rectangular bags can be unzipped and then the two bags can be joined together to make one large queen-size sleeping bag for two people. When two people share the same bag it helps to provide warmth, companionship, and a feeling of security. Just knowing that you could reach over and touch someone during the night relieves a lot of the psychological stress of wilderness living.

Shelter Size: If you use plywood to construct the exterior walls of your home, and paneling to dress up the interior of your home, then you should probably build a home that is some multiple of four-feet in each direction since that would be the size of your building materials. The roof of your home should be a simple flat shed roof sloped from the rear of your home towards the front so that rainwater will run off your roof onto the ground in front of your home. This would be the case even though you have between four to six-inches of dirt on your roof because you want the water to flow down the slope and fall off the front of your roof. Plant grass seed in the dirt on your roof and you will have a sod roof.

Insulation: If possible you should install insulation inside your front wall because it will be the wall where you will have most of your heat loss. If possible, also insulate your ceiling. You could also add insulation inside your other three exterior walls and then you would have a very comfortable home. A complete list of building supplies with their cost is in Appendix One of this book.

Shelter Exterior: Instead of plastic or aluminum siding I suggest heavy-duty 10 mil thick tarps as the outside moisture and insect barrier around your entire shelter. The reason is because these tarps are a continuous sheet of thick fabric and therefore moisture and insects cannot easily get past them into your shelter. On the other hand, plastic or aluminum siding has very tiny gaps between each piece of sheeting and water and insects can more easily get through those gaps and into your home. The best option, but the most expensive, would be to install aluminum siding first and then cover them with 10 mil thick tarps on the exterior of your new home.

The Seasons and Shelter Temperature: If you build an underground shelter then about eight months of every year will be very pleasant months. The temperature inside your underground shelter will fluctuate between 50°F to 60°F (or 10°C to 16°C) all year. In the warm months you can open your windows just a little to let warm outside air into your shelter until it reaches a temperature that is agreeable to you. Most people like an indoor temperature between 65°F to 75°F (or 18°C to 24°C). This will be relatively easy to achieve during warm weather by opening and closing your windows based on the current indoor and outdoor temperatures.

However, you need to be prepared for cold winters. The temperature inside your shelter will stay at approximately 50°F (or 10°C) during the cold winter months. The reason is because the mass of the earth around your shelter will be keeping the temperature inside your shelter at the same temperature as the surrounding earth. The primary heat loss will be through your ceiling and your front wall and if you have those areas reasonably well insulated then your heat loss will only be through your windows and door. If you put a think heavy blanket over each window and door then you can minimize this heat loss at night. However, you will need to remove the blanket from at least one of your windows during the day to allow the natural daylight into your shelter. During sunny winter days the sun coming through your windows will gradually increase the temperature inside your shelter and make it more comfortable even though the outside air temperature may be below freezing. But on overcast winter days the temperature inside your shelter will be about the same as the surrounding earth. Your shelter should be below the frost line so it will be cool but not freezing inside your shelter. Low winter temperatures are a fact of life that you need to be intellectually and emotionally ready to accept.

If you build a fire inside your shelter then you will need to vent the shelter to the outside to allow the smoke to escape. And the smoke will drift on the wind and lead anyone who is downwind directly to your new home. Therefore the safest and the most efficient way to stay warm during the cold winter months is to dress warm while you are inside your new shelter. You should wear several layers of clothing, two pair of socks, a pair of soft thin flexible comfortable gloves, and a cap with ear flaps or a ski mask.

Chapter Thirteen

Equipment Waist Belts and Waist Packs

I have found it useful to carry some of my wilderness survival items in two different locations:

1. on a waist belt, and
2. in a waist pack.

Let's look at each of these two storage options one-at-a-time.

Equipment Waist Belt

It is not practical to attach items to the belt that holds your pants up. If you do then those items will pull your pants down as you walk and rub blisters on your hips. Not only is this painful but it is also very annoying. Therefore let your normal belt (or suspenders) hold your pants up. Then use another belt for your knife, pistol, and canteen but don't thread that belt through the belt loops on your pants. I prefer a belt canteen to a shoulder strap canteen because the shoulder strap wears a blister on your shoulder in a very short time, even if you keep changing shoulders.

Now let's look at the weight of the items on my waist belt. The weights in the tables that follow are in decimal pounds which makes addition and subtraction relatively easy. I chose that method instead of pounds and ounces so you would not have to convert ounces to pounds to find the total weight.

Items on an Equipment Waist Belt

Item Description	Pounds
Empty Military Carry-All Belt	0.70
Canteen with 30 Ounces of Water	2.32
Hunting Knife in Sheath	1.25
Loaded Pistol in Holster	3.11
Waist Belt Total	**7.38**

Waist Pack

I carry the items I use most often in a waist pack. The reason is because I want those items with me when I explore the wilderness area near my shelter and when I go hunting or fishing.

These are the items I carry in my waist pack:

Small Waist Pack

Item Description	Pounds
Empty Waist Pack	0.58
Compass	0.14
Pocket Watch and Extra Battery	0.18
Flashlight and a Headband (2AA Batteries)	0.21
One Butane Lighter (3000 Lights)	0.03
60 Matches in Waterproof Container	0.06
Magnesium Fire Starter (Emergency)	0.12
Metal Whistle (Communication)	0.03

Fishing Tackle Kit (No Rod or Reel)	0.36
30 Feet of 20 Gauge Snare Wire	0.05
Rubber Gloves (Skinning Game, Cleaning Fish)	0.16
Leather Work Gloves	0.47
4 One-Gallon Freezer Bags (Fish, Game, Plants)	0.12
10 Kleenex Tissues (Emergency Toilet Tissue)	0.05
First Aid Kit (7" X 4.5" X 1.5")	0.80
Snake Bite Kit	0.08
Area Map	0.09
Mosquito Head Net	0.09
Plastic Rain Poncho (Cost $1)	0.15
Waist Pack Total	**3.77**

Fire is an absolute necessity to have in the wilderness. You can heat a meal, or cook raw food, or use it to get warm by, or keep wild animals at a distance at night. And it is a moral booster like nothing I know when you are alone. If you should get lost then you can start a signal fire and create a smoke column to draw rescuers to your location. Therefore I always have one butane lighter with me and some matches in an empty waterproof match container. Some matches fit really well in those plastic containers but you may have to trim a little off the wooden end of some of the longer matches to get them to fit. If you don't have the strike-anywhere matches then cut the striking strip off the match carton and put it inside the waterproof container with your matches. But be careful that the match heads don't rub against the striking surface. You can wrap the striking surface in clear plastic wrap to protect it and keep it dry. The Magnesium Fire Starter is for emergency backup use only. If for some reason I need a fire immediately and all the nearby tinder is frozen or soaking wet then I can shave some magnesium of the magnesium block and light it. The magnesium will burn extremely hot for a short period of time and it will usually dry out my small thin sticks so they will catch fire. If I can get my thin sticks burning then I can usually get my larger sticks burning.

I carry those small 10-tissue packs of Kleenex with me for use as emergency toilet paper. The Kleenex is a true morale builder just knowing I have it in reserve if I should need it. However, if I don't have any tissues then I use several big green leaves instead of toilet tissue.

I don't carry a fishing rod or reel but I do have a reasonable selection of fishing tackle items, such as assorted fishing hooks in various sizes, swivels, sinkers, fishing line, some plastic worms and grubs, lots of flies, and a 1/8 ounce white rooster-tail fishing lure.

I carry a pocket watch in my waist pack instead of a wrist watch. It doesn't get banged up as much as a wrist watch when I am in the wilderness. My pocket watch is not a wind-up unit but the battery is new and it should last for 3 or 4 years. And it has hands on it so I can tell which direction is south if the sun is visible. Just point the short hour-hand at the sun and south will be half-way between the short hour-hand and the number 12 on the watch. Works consistently on normal time but not daylight saving time. (During daylight saving time I use the number 1 instead of the 12.) Plus you have to be north of the equator. (Note: If you are south of the equator then point the number 12 at the sun and north will be half-way between the number 12 and the short hour-hand.)

I have a nice GPS receiver and I really love it. It does all the thinking for me. On the other hand, I don't use it anymore. I found it is too much like the hand calculator I began using in the 1970s. I have to pause and think in order to do math in my head nowadays. I am ashamed of myself. That's why I don't use my GPS unit anymore. If I relied on it then I would not pay as much attention to my surroundings. If my GPS unit stopped working for any reason then I would be lost in the wilderness. But if I don't use it and I rely upon my powers of observation instead then I can continue to find my way back from anywhere I go.

Chapter Fourteen

Some Additional Issues about Long-Term Wilderness Survival

Recycling: After you arrive at your new wilderness location and before you throw anything away you should think about how you could use that item in the future. Glass and plastic bottles with screw on lids should be washed and saved. Empty tin cans and their lids should be washed and saved (may be melted and formed into something useful). Any type of paper or cardboard should be saved (future tinder for a fire). Almost everything you bring with you into the wilderness will have value after you have emptied the original container. If you think about this ahead of time then you will be able to better select your food items based on the type of container the food is sold in. But there are exceptions. For example, canned soda has a longer shelf life than soda in a plastic bottle. Generally canned food has a longer shelf life than food in glass or plastic containers. Therefore it would probably be a good idea to buy at least three-fourths of your food in cans instead of plastic or glass.

When you first arrive in the wilderness the most important issues that must be immediately addressed are safety, shelter, water, food, personal hygiene, and entertainment. If you can adequately address these issues then you will have a much better chance for long-term wilderness survival.

Safety: An area that is not beside a road, or lake, or creek, or other source of water is usually a safer place to be. The lack of a nearby road or a reliable water source will usually mean the area will not be visited very often by predators (animals or humans). You can use your tarps to capture rainwater on a regular basis and replenish your water supplies. If there is a serious drought then you can always hike to the nearest water source and replenish your water supply. Always boil and then filter any ground water you collect.

Be extremely careful all the time. Emergency professional medical care is not a few minutes away. A normal accident could leave you crippled for life. A serious accident could cost you your life. Although you are probably already a very careful person, you need to be even more careful in the wilderness.

If you do not already wear normal eyeglasses then it would be a good idea to always wear safety glasses when you are awake regardless of what you are doing. If you can preserve your eyesight then you have a chance to survive in the wilderness. If you lose your eyesight then you have no chance at all.

Be very careful whenever you are using your knife or any of your tools. If you are right handed then your left hand will suffer the most damage. Your right hand will be holding the knife, or the hatchet, or the hammer while your left hand holds the nail, or wood, or fish. Therefore you left hand will get hit or get cut by the instrument in your right hand. Although this is obvious most people don't think about it even after they hurt their left hand. However, you can help to eliminate these accidents by thinking about it ahead of time.

Also practice extreme caution you are handling any of your firearms, and while hunting, and when you check your animal traps. Any wild animal could cause you serious injury with its teeth or its claws or nails and that includes a helpless looking little rabbit. (Note: A rabbit put long deep scratches in my left forearm.)

Also practice good safety rules when eating or drinking anything. If you are eating canned food that needs to be heated then heat the food to the boiling point in order to neutralize any harmful microorganisms that might be in the food. This is a simple safety precaution but it could easily save your life. If you do not have a quality water filter then you should always boil all your drinking and cooking water to prevent serious health problems.

Another safety issue is human predators. There are no law enforcement officers in the wilderness to protect you, or to bring a criminal to justice if you are attacked and murdered in the wilderness. Therefore you will need to be constantly alert for the presence of other human beings. If possible you should try to remain out-of-sight so you will not be seen by the causal hunter or hiker, or by the hardened criminal who is trying to escape justice by hiding in the wilderness. Your safety is a serious issue and you should make every effort to keep the location of your new shelter a secret to the extent possible. One way you can do this is to try to minimize the noise level during construction. This will be the one period of time where you will be making a lot of noise and that noise could easily attract the attention of anyone within a mile of your building site. Think about that before you make too much "unnecessary noise" during construction.

Protection: Always wear your sidearm and your hunting knife on an extra belt around your waist (except when you are asleep). When you are in the wilderness you never know who or what might wander into your camp and you need to have the means to protect yourself all the time. And always have a loaded rifle somewhere nearby that you can get to in just a few seconds if the need should arise.

Emergency Electric Heat: Build a tent over your bed using blankets or extra sleeping bags and then turn on a single 40 or 60 or 75 watt lamp that is on the table beside your bed but that is also inside your tent shelter but not touching your tent. An incandescent bulb (not an energy efficient or florescent bulb) will provide light and heat. It will not consume oxygen. It will not generate a lot of heat but the temperature inside your temporary tent shelter will increase by several degrees and if you put your hands close to the light bulb you will be able to feel its heat.

Nuclear War: In the event of a nuclear war, or a nuclear power plant meltdown, all the water everywhere will be contaminated with tiny radioactive particles. At the time of the blast (or nuclear plant meltdown) some of those tiny particles will make their way into the upper atmosphere and they will gradually move around the earth. Therefore the ground water everywhere will eventually become contaminated. The worst contamination will be closer to the blast and lower levels of contamination will be further away. The ground water will remain contaminated for many years. However, with the passage of time rainwater will contain less and less radioactive particles as the rain washes the radioactive particles out of the atmosphere and deposits them on the earth where they will eventually make their way into the ground water everywhere. Therefore, if there is a nuclear war then you should filter all your rainwater through a good water filter before you bring it into your shelter. This will keep the radioactive particles out of your shelter, and out of your body, and off your body, dishes, and clothing. The possibility of a nuclear war is one of the reasons you should always have at least a 60-day supply of drinking water inside your shelter. You should also have some type of face mask to wear when venturing outside your shelter after an explosion. A professional quality gas mask would be great but a simple doctor's or painter's face mask would also be fine. If you don't have any type of face mask then tie a cloth around your mouth and nose and breath through the cloth when you go outside. Then leave the cloth outside when you enter your shelter. You should also leave your contaminated outer clothing and your water filter outside your shelter to avoid bringing radioactive particles into your shelter that might have become attached to your exterior clothing and hat. If possible do not leave your shelter for at least six-weeks after a nearby nuclear explosion.

Personal Hygiene: In addition to all your normal personal hygiene activities you should also wash and dry your feet every night after a long day of hard work. Put on a clean pair of socks every night before going to bed and then wear those socks the next day.

Entertainment: If possible invest at least ten-minutes each day doing something you really enjoy. For example, after the sun goes down you could listen to your radio, or watch a movie on your portable computer with a DVD player, or read a book if you have battery powered lights (12-volt ceiling lights or a battery powered L.E.D. lantern).

Chapter Fifteen

Some Practical Strategies for Long-Term Wilderness Survival

You will need to decide whether or not you wish to become invisible in the wilderness. If you do desire to disappear then there are three things you must be very careful about: **light, sound, and smell**.

Light: After dark it is very easy to see any type of light including a campfire. However, if you are living inside an underground shelter then you could simply put a very thick blanket over your windows before you turn on your lantern (or 12-volt ceiling lights). Therefore light will probably not be a problem for you.

You can also avoid visual detection during the day by wearing dark color or camouflage clothing instead of whites or light colored clothing. Obviously dark clothing would not be appropriate in a snow covered environment but hopefully you will not need to venture very far outdoors during the bitter cold winter months when there is snow on the ground. However, in that situation white outer clothing would be advantageous. If you settle in one of the warmer central or southern climates then the number of days the ground will be covered with snow each winter should be few and far between.

Sound: Sound carries a long distance in the wilderness and it is relatively easy to differentiate natural forest sounds from the sounds of mankind. However, if you are living in an underground shelter then you will not have to worry about any stray sounds you make inside your shelter being heard by someone or some animal outside your shelter This means you could laugh, cough, or snore and nobody outside your shelter would hear you. However, whenever you are outside your shelter you will need to be careful to not make any unnecessary noises. This includes chopping wood, sawing wood, or talking. These sounds travel a good distance and they are easy to identify as human sounds. Gunfire also travels a really long ways so you need to think about this before you go hunting. If you are able to capture most of your wild game using traps then this will not be a problem. You could also shoot small game animals with an air rifle that shoots hollow point or pointed pellets. However, if you occasionally take a big game animal with a hunting rifle then you will need to be aware that your shot will be heard a long distance away. Generally it is very, very difficult to accurately determine the direction from which a single shot was fired because the sound was unexpected. However, the moment a single shot is heard everyone becomes instantly alert for another shot. The second (or third) shot is the one that enables a person to clearly identify the exact direction from which the shooting is being done. Therefore, if possible, always try to bring a big game animal down with a single shot. If you use two or three shots then be aware that anyone within a mile or more will have heard you and that person will have a pretty good idea exactly where you are. Therefore you should process the animal as quickly and safely as possible and remain extremely alert while you are completing your task.

On the other hand, if you fire two or three very quick shots using a semi-automatic rifle then the sounds of those shots will be very close together and it will be almost the same as the unexpected sound of a single shot. However, if you pause one or two seconds between shots then those shots will not be unexpected and anyone who is carefully listening for a follow-up shot will be able to more accurately determine your position.

Smell: Smell will probably be the most challenging thing for you to control. The wind will carry smells a very long ways. The most common smells are smoke and food. Freshly baked bread and freshly brewed coffee have very strong smells and they can be easily detected on the wind from a very long distance away. Almost anyone can follow a smell back to its source even if they have not had any training or previous experience in this type of tracking. All you have to do is determine which direction the wind is blowing from and then walk into the wind. You then turn right or left depending on which direction has the strongest smell. If you turn right for a short ways and the smell becomes fainter then you immediately turn left into the wind and travel in that direction for a while. By walking into the wind and making a zigzag path from left to right, and then right to left, you will stay on the scent trail until you eventually reach its source.

If you are living in an underground shelter then you will not need to build a fire for warmth so you will not be sending out a smoke signal that would lead someone to your location. If you use a microwave oven for heating your food then any food smells will be trapped inside your shelter. As the cooking odors gradually fade you can open your door briefly to let in fresh air and then close the door again. This would let the cooking smells out of your shelter but it would be a one-time short-duration smell as opposed to a continuous long-duration smell that could be easily followed back to your shelter. A simple example would be the smell of a single match that soon goes out versus a small campfire that is left burning for several hours. The smell from a single match is very powerful but it quickly dissipates into the surrounding air until it becomes unnoticeable. However, a small campfire that is left burning for a long time emits a continuous column of smoke that can be easily tracked back to the campfire itself. Therefore when you need to cook over a campfire you should use seasoned dry wood that is almost smokeless. And you should use covered skillets or a Dutch Oven to keep your cooking odors trapped inside your cookware except for the brief moments when you lift the lid to stir your food or to check on your food.

Garden Fertilizer and Compost

Your future gardening success will be in direct proportion to how you manage your garden soil. Each time you harvest your garden crops you will be depleting the natural minerals and nutrients in the soil. Those minerals and nutrients must be replenished or you will exhaust your soil and nothing of value will grow there. If you rotate your crops and grow a complimentary crop in the soil the second year then you can help minimize the depletion of nutrients in your garden soil. In a wilderness environment the easiest way to enrich your soil is to have your own compost pile near your garden area. Any vegetable or plant material that will properly decay could be added to your compost pile. You can also safely recycle your own human waste if you carefully follow the procedure discussed in the compost chapter of my book **"Grandpappy's Survival Manual for Hard Times."**

Self-Sufficiency

Resist the temptation to become 100% self-sufficient in everything. However, you will need to become self-sufficient in providing your own emergency first aid and you will need to provide most of your own food. But you will still need to purchase salt and peppercorns. Tone's brand black peppercorns can be purchased in medium size plastic jars at a very reasonable price at Walmart. (Note: You will also need a pepper grinder.) Morton canning and pickling salt can also be purchased in four-pound boxes at a very reasonable price at Walmart.

Herbal Remedies

Resist the temptation to become a herbalist. Do not trust your health and healing to medicinal herbs. There is nothing wrong with medicinal herbs as a last-resort option but in most cases medicinal herbs are not as effective as currently available medications. My opinion is based on several years of experience growing my own medicinal herbs. However you do not need to take my advice on this subject -- you can learn for yourself if you wish. But my recommendation is that you should have a reasonable supply of over-the-counter medicines (such as aspirin, ibuprofen, and anti-diarrhea tablets) that you can purchase at any drug store or Walmart. On the other hand, it is okay to grow cooking and seasoning herbs, such as oregano, if you wish.

Farm Livestock

Resist the temptation to add any type of farm livestock to your wilderness retreat. This includes chickens, rabbits, goats, cows, horses, and dogs. These animals will take more of your time and effort to manage than the benefits they give back in return. They will also make noise that will attract attention to your wilderness retreat. You will be much better off purchasing instant powdered dry milk, and replenishing your meat supplies by fishing, trapping, and hunting. Ride a bicycle into town. And learn how to live without fresh eggs.

Target Practice

Conserve your ammunition. Do not waste your ammunition killing paper targets. Your ammunition has much more valuable uses than shooting pretty holes in pieces of paper. However, you will need to zero the scope on each of your rifles so you can hit a wild game animal at the exact spot it appears in the center crosshairs of your scope. Centering your rifle scope is not a waste of ammunition -- it is an investment. If your rifle bullet does not go exactly where you expect it to go then you could miss the wild game animal and you would lose:

1. your bullet, and
2. the meat, and
3. the hide of the animal.

Therefore a properly centered rifle scope is a necessity and not a convenience. However, all other normal target practice is primarily for entertainment. When you cannot easily replenish your ammunition supplies this is one entertainment activity that you should avoid. In the long-run your ammunition will provide you with safety and food and these are two things you do not want to give up for the sake of a temporary entertainment activity that you intellectually justify as "necessary self-defense practice."

Blacksmithing

Do not try to become a professional blacksmith. However, you should learn how to make simple repairs on the items you already own. But you should not try to forge your own steel. It is much more practical to simply buy a few good stainless steel knives and hand tools and garden tools instead of trying to fabricate your own tools from scratch. This would not be a productive way to invest your time. And the product you eventually construct will probably not meet your minimum quality requirements.

Spinning Wheels and Weaving Looms

Do not try to make your own thread or cloth. The amount of time and effort you will invest in these simple activities will not yield a product that is acceptable to you in terms of quality. Instead, when you occasionally visit a nearby town you should purchase professionally manufactured thread and cloth to make new clothes, or you should buy new or used clothes. And you should have previously purchased enough hand sewing needles and thread at the very beginning of your adventure to do all your clothing repairs for the rest of your life. You should also have purchased a few basic pieces of cloth so you can make some basic replacement clothing items as the need arises. Or you can make replacement clothing and footwear from deer skins that you process into buckskins.

Periodically Replenishing Your Supplies

You will need to practice skinning deer and other animals so you can salvage and brain tan their hides. You will also need to learn how to convert those hides into buckskins.

For example, if you have some nice deer skins, beaver skins, or a nice bear skin, then you could sell or trade those hides at a "flea market" for the things you can't make, such as salt, pepper, toilet tissue, and ammunition.

However, you should have taken enough of these non-replaceable necessities with you into the wilderness to last for at least 3 or 4 or 5 years. This will give you time to develop your skills so you can successfully make things, or grow things, that you could eventually sell or trade at a "flea market" or a temporary "road side stand."

A Companion of the Opposite Sex

You will not be able to fully understand and appreciate the meaning of the word "lonely" until you are completely alone in the middle of nowhere for an extended period of time. This unpleasant feeling can be avoided if there is at least one other person with you (for example, your spouse and your children). However, your companion must be someone you can trust to stay with you after the novelty of the adventure wears off and the harsh realities of wilderness survival become painfully obvious. Before entering the wilderness men and women will both have a very positive attitude about the adventure and their ability to "tough it out." Unfortunately most of them will simply give up after a short exposure to the boredom, hard work, and isolation of wilderness living.

Giving Up

The eight major issues that will make the average person want to give up are:

1. the cold winters when the temperature inside your shelter stays at approximately 50°F (10°C) all winter,
2. the amount of work required to cook most of your food from scratch (no McDonalds or frozen pizzas),
3. the amount of work required to safely dispose of your own human waste,
4. the amount of work required to provide your own safe drinking water,
5. not having hot water instantly on demand 24-hours per day 365 days per year,
6. not having a dishwasher or washing machine and having to do your dishes and laundry by hand,
7. not having several hundred cable TV channels for entertainment, and
8. not being able to go shopping whenever you get the urge to buy something.

These are things most of us take for granted and when they are no longer available (or easily taken care of) then it is only natural for a person to complain. But complaining will not do any good. You will need to live without these things the same way your ancestors did. However, if you know ahead of time that these issues will never be acceptable to you then you should not venture into the wilderness to begin with.

Conclusion

Finally, you will need to learn more about your Creator. Fortunately He has provided you with an accurate detailed reference manual -- The Holy Bible. Read it every day. Pray every day for guidance and protection.

Appendix One

Basic Construction Plans for an Underground Home

You have a variety of options for your future wilderness shelter. One of those options is a simple two-room underground home built into the side of a hill. The pages in Appendix One contain information about this type of wilderness shelter.

An Underground Home Built From Lumber and Plywood

On page 63 of Appendix One there is a simple sketch of the interior of a two-room wilderness home. This home could be constructed from lumber and other common building materials for about $4,100.00 (United States Dollars).

A complete bill-of-materials with prices is on pages 61 and 62. The material list is divided into several sections. Following is an explanation of each of those sections:

Basic Minimum Supplies: These few items would be highly desirable even if you build a "cabin cave" using pine trees that grow on your land.

Basic Foundation, Frame, and Exterior Structure: These items would enable you to provide a firm foundation for your home, and build a strong frame structure, and add four exterior walls, a roof, and a floor.

Optional Interior Walls, Ceiling, and Floor: These items would allow you to add interior walls so your new home would more closely resemble the type of home you have probably lived in most of your life.

Optional Roof Covering: These items would improve the quality of your exterior roof.

The remaining items are self-explanatory. These are items you could add to the interior of your home to improve its usefulness and functionality.

Optional Kitchen Cabinets

Construction Suggestions

The floor of the shelter should be at least 8-feet above the bottom of the hill to avoid future flooding problems.

The front wall that contains your door and two windows should be facing south, southeast, or southwest (in the northern hemisphere). These windows will be your only source of natural sunlight and moonlight and they need to be strategically located so they will capture as much of this light as possible.

If you decide to build a shelter with 7-feet tall interior walls then your should leave your lumber 8-feet tall on the rear wall of the shelter. The other walls should be 7-feet tall so the inside of your shelter will resemble the normal rectangular living space to which you are accustomed. If you add an additional six-inches of height to the exterior rear wall then this will provide a slope of 18-inches from rear to front for a flat shed type roof that slopes from the rear to the front of your shelter to facilitate water runoff when it rains.

After you finish digging the hole into the side of the hill, and before you begin construction, you should place a heavy-duty 10 mil tarp on the ground below your future floor to minimize future insect and termite problems.

The exterior walls, whether pine logs or plywood, should be coated with a weather-proof waterproof seal to prevent future moisture problems. Then a heavy-duty 10 mil tarp should be placed on all the outside surfaces before you pile dirt against your walls. This will help to keep moisture off the lumber and out of your home. If you can afford it you should install plastic or aluminum siding between your outside plywood walls and your 10 mil tarps. If you can afford it, normal wall insulation should be installed in your front wall and in your ceiling to prevent heat loss during the cold winter months. If you can afford it you can also put insulation in your other three walls.

Follow normal construction guidelines while building your home. For example, after you have the frame and outside walls installed, then you should do all the electrical wiring before you add the interior walls. If you put down your sheet of vinyl linoleum flooring before you install the center wall and the plywood interior walls then the edges of your vinyl floor will be kept tight against the floor by the plywood walls and the center wall.

Bill of Materials For Underground Shelter Built into the Side of a Hill
Requires Knowledge of Basic Home Construction Techniques
(Layout Sketch on Page 63) (20 Feet Long by 12 Feet Wide by 8 Feet Tall) (Prices January 1, 2009)

Description	Quantity	Price Each	Total Price
Basic Minimum Supplies:			
36" Steel Door Complete (Hung Right Hand - RH)	1	116.00	116.00
Front Door Lock with Separate Dead Bolt Lock (RH)	1	30.97	30.97
Four Extra Front Door Keys Made (Bury 2 Keys Outside)	4	1.25	5.00
36" Window in Frame With Screen and Storm	2	134.00	268.00
10'x20' Tarp 10 Mil (12'x20' = $21.88) (9'x12'=$12.88)	4	19.97	79.88
16'x30' Tarp 10 Mil	2	34.97	69.94
5-Gallon Water Seal For Exterior Wood	2	39.00	78.00
1-Gallon Dark Exterior Paint for Door and Windows	1	22.44	22.44
Disposable Paint Brushes (2" wide and 4" wide)	4	1.00	4.00
Sub-Total			**674.23**
Basic Foundation, Frame, and Exterior Structure:			
8"x8"x16" Concrete Blocks	57	1.27	72.39
Mortar Mix (60 Pound Bag)	14	3.98	55.72
2"x4"x8' Lumber (Exterior Walls)	56	2.23	124.88
2"x4"x8' Lumber (One Interior Center Wall)	10	2.23	22.30
2"x6"x12' Lumber (Floor and Foundation)	44	6.26	275.44
2"x6"x16' Lumber (Shed Type Sloped Roof)	18	7.98	143.64
30 Pounds 3" Interior Building Nails	1	36.89	36.89
30 Pounds 1.5" Interior Building Nails	1	36.89	36.89
4'x8'x28/32" Plywood (Exterior Walls)	16	17.87	285.92
4'x8'x28/32" Plywood (Exterior Roof)	10	17.87	178.70
4'x8'x28/32" Plywood (Interior Floor)	8	19.87	158.96
Sub-Total			**1,391.73**
Optional Interior Walls, Ceiling, and Flooring:			
5 Pounds 3/4" Interior Nails	1	9.98	9.98
3.5"x15"x32" Wall Insulation R-13 (Front Wall Only)	4	9.98	39.92
3.5"x15"x32" Wall Insulation R-13 (Ceiling Only)	5	9.98	49.90
20'x25' Black Plastic Sheeting 3 Mil (Walls, Ceiling)	4	26.54	106.16
4'x8' Paneling (Interior Walls)	16	14.78	236.48
4'x8' Paneling (One Interior Center Wall)	6	14.78	88.68
4’x8’ Paneling (Interior Ceiling)	9	14.78	133.02
12'x20' Continuous Sheet Roll of Floor Linoleum	1	225.60	225.60
1-Gallon Can of Glue for Linoleum Floor	1	15.90	15.90
Sub-Total			**905.64**
Optional Roof Covering:			
Roof Asphalt Paper (102 sq. ft. per roll)	3	29.35	88.05
Roof Shingles 30 Year (33.3 sq. ft. per bundle)	9	26.50	238.50
5 Pounds 7/8" Roofing Nails	1	9.98	9.98
Sub-Total			**336.53**
Optional 6-Foot Storage Shelf Unit:			
3/4"x16"x72" Shelf Lumber (5 Shelves)	5	24.31	121.55
3/4"x16"x96" Shelf Lumber (2 End Pieces)	2	32.62	65.24
Sub-Total			**186.79**
Optional 6-Foot Kitchen Counter and Cabinets:			
72" Formica Counter Top	1	66.00	66.00
36"x34" Lower Cabinet (2 Doors, 2 Drawers Each)	2	159.00	318.00
36"x18" Upper Cabinet (2 Doors Each)	2	105.00	210.00
Cabinet Hardware Handles (8 Doors, 4 Drawers)	12	1.97	23.64
Sub-Total			**617.64**
Grand Total for the Entire Shelter			**$4,112.58**

Optional Items For Underground Shelter Built into the Side of a Hill

Description	Quantity	Price Each	Total Price
Outdoor Propane Tank:			
100-Pound Propane Tank Filled with Propane	1	239.00	239.00
Propane Pressure Regular for a 100-Pound Tank	1	49.00	49.00
Flexible 12-foot Propane Hose (Tank to Heater)	1	29.93	29.93
Sub-Total			**317.93**
Note: Select One of the Following Heaters:			
Buddy Indoor Safe Propane Heater (no fan)	1	79.00	
Blue Flame Safe No Vent Propane Heater	1	129.00	129.00
Infrared Safe No Vent Propane Heater	1	185.00	
Sub-Total		**Average =**	**129.00**
Optional: Select One of the Following Ovens:			
Maytag Propane 22-inch R.V. Oven and Stove	1	382.00	?
Free Standing Propane Oven and Stove	1	550.00	?
Minimum Solar Power Equipment:			
60-Watt Solar Panel (or four 30-Watt Solar Panels)	2	332.98	665.96
Pro-Star 30 Amp PS-30 Solar Charge Controller	1	137.41	137.41
12-Volt Marine or Golf Cart Deep Cycle Battery	4	78.94	315.76
2-Gauge or 4-Gauge Battery Cables (Black 15" to 24")	3	5.98	17.94
2-Gauge or 4-Gauge Battery Cables (Red 15" to 24")	3	5.98	17.94
Sub-Total			**1,155.01**
12-Volt DC Electrical Power:			
50-Feet 14-2 Gauge Electrical Wire (2 wires)	4	23.96	95.84
12-Volt Camper Ceiling Light (L.E.D. if available)	6	7.89	47.34
12-Volt Wall Switch Complete (one per ceiling light)	6	5.29	31.74
Replacement 12-Volt Light Bulbs (auto parts store)	12	1.29	15.48
12-Volt Cigarette Lighter Outlet	6	9.98	57.88
Sub-Total			**248.28**
120-Volt AC Electrical Power:			
Inverter 2,000 Watt Continuous / 4,000 Watt Peak	1	211.92	211.92
50-Feet 14-3 Gauge Electrical Wire (3 Wires)	4	32.97	131.88
120-Volt Electrical Outlet (box, outlet, face plate)	6	4.36	26.16
Package of four 60-watt incandescent light bulbs	2	1.89	3.78
Sub-Total			**383.74**
700-Watt Small Microwave Oven (AC Power)	1	62.00	62.00

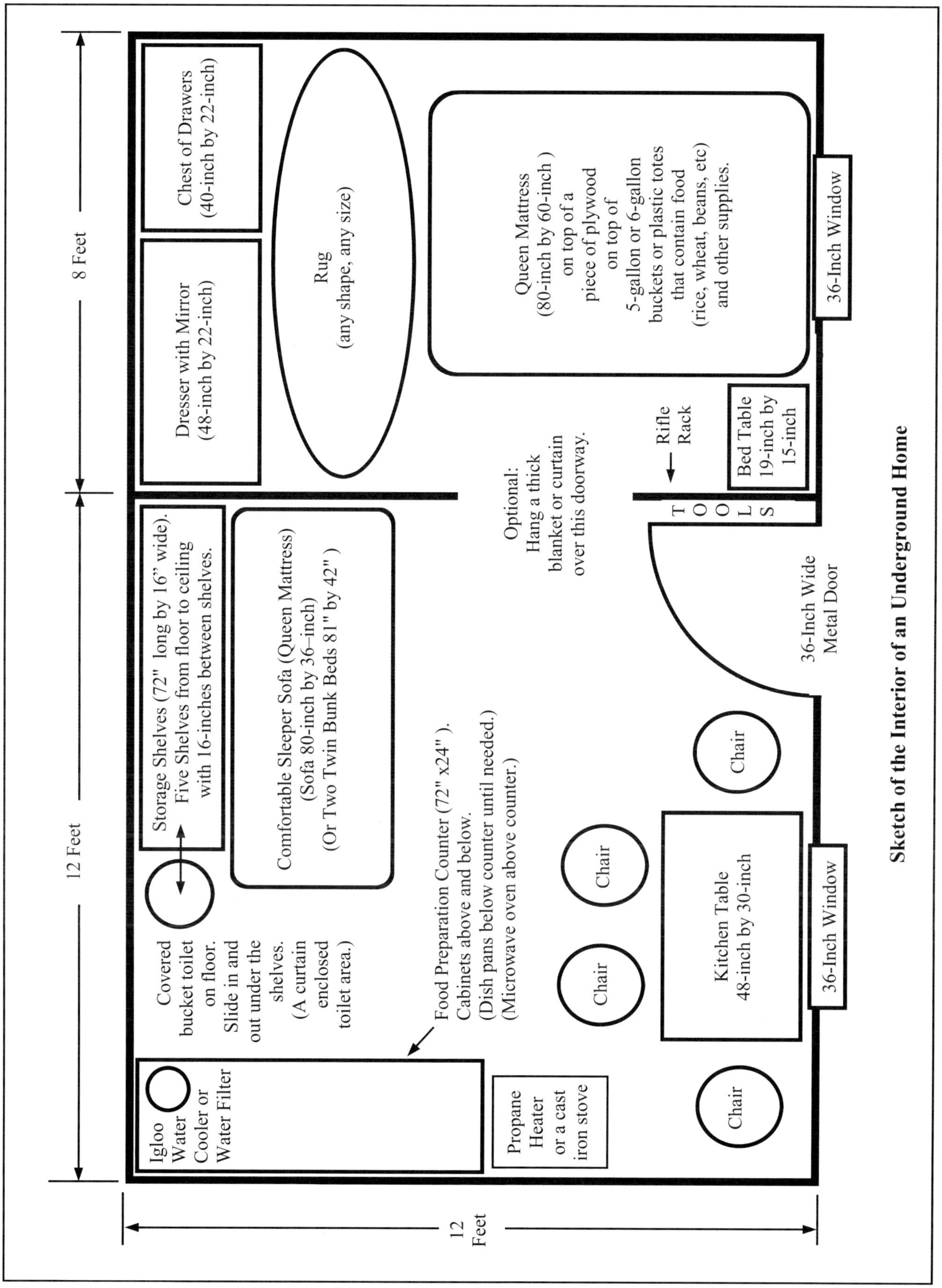

Sketch of the Interior of an Underground Home

Appendix Two

Practical Wilderness Survival Information Available on the Internet

The Following Web Sites Contain Practical and Useful Information.
Download the Files and Print Hard Copies and Store Them in
a Three-Ring Binder For Future Reference.

How to Make Moccasins:
http://jumaka.com/moccasins/MakingMoccasins/MoccasinInstructions.htm

How to Make Tire Sandals:
http://www.hollowtop.com/sandals.htm

How to Darn Socks:
http://www.ehow.com/how_648_darn-sock.html?ref=fuel
http://www3.telus.net/findNchoose/Darning.html
http://www.wikihow.com/Darn-a-Sock

Solar Oven Instructions:
http://www.cookwiththesun.com/wood-box.html
http://www.oksolar.com/samples/sunoven1.html

Block and Tackle Explanation:
http://www.howstuffworks.com/pulley.htm

Smoke House Information:
http://www.endtimesreport.com/smokehouse.html
http://www.wedlinydomowe.com/smokehouse-meat-smokers.htm

Root Cellar Information:
http://www.tribwatch.com/rootcell.htm

Ranger Rules of Engagement:
http://en.wikipedia.org/wiki/Rangers_Standing_Orders

The Michigan State Trapper's Guide:
http://www.michigan.gov/documents/dnr/MI_Trapper_Education_Manual_82307_206561_7.pdf

The Ten Bushcraft Books:
http://tions.net/CA256EA900408BD5/vwWWW/outdoor~03~000

Appendix Three

Links to Free Survival Information on Grandpappy's Web Site

Safe Temporary Shelter -- https://www.grandpappy.org/wtempshe.htm
Cabin Cave -- https://www.grandpappy.org/wcave.htm
Bolt-Action or Semi-Automatic Firearms -- https://www.grandpappy.org/fcompare.htm
How to Hit the Target Bull's-eye -- https://www.grandpappy.org/hbullsey.htm
Laser Sights -- https://www.grandpappy.org/flaser.htm
Homemade Bow and Arrows -- https://www.grandpappy.org/wbow.htm
Gillnets -- https://www.grandpappy.org/wgillnet.htm
Steel Traps and Snares -- https://www.grandpappy.org/wsnares.htm
Basics of Shortwave Radio -- https://www.grandpappy.org/htshortw.htm
Flashlights -- https://www.grandpappy.org/htflashl.htm
Solar Power -- https://www.grandpappy.org/htsolar.thm
Rechargeable Batteries and Solar Battery Chargers -- https://www.grandpappy.org/htcharge.htm
Gardening Information -- https://www.grandpappy.org/indexgar.htm
How to Grow Fruit Trees from Seed -- https://www.grandpappy.org/gfruitse.htm
How to Convert Human Waste into a Safe Compost -- https://www.grandpappy.org/gcompost.htm
One-Year Food Supply -- https://www.grandpappy.org/hfoodaff.htm
Acorns -- https://www.grandpappy.org/racorns.htm
Wild Game Recipes -- https://www.grandpappy.org/rgame.htm
Squirrel Recipes -- https://www.grandpappy.org/rsquirre.htm
Fish Recipes -- https://www.grandpappy.org/rfish.htm
Kudzu Recipes -- https://www.grandpappy.org/rkudzu.htm
Poke Sallet Recipes -- https://www.grandpappy.org/rpoke.htm
Meat Jerky Recipe -- https://www.grandpappy.org/hsksmoke.htm
Grandpappy's Pemmican Recipe -- https://www.grandpappy.org/rpemmica.htm
Cast Iron Cookware -- https://www.grandpappy.org/hskcasti.htm
Water -- https://www.grandpappy.org/hwater.htm
Toilet Tissue -- https://www.grandpappy.org/htissue.htm
Grandpappy's Homemade Soap -- https://www.grandpappy.org/hsksoap.htm

Appendix Four

Recommended Books for Basic Wilderness Survival

1. **SAS Survival Handbook**, Collins Gem, John "Lofty" Wiseman, Highly Recommended.
2. **Survival Gunsmithing**, J.B. Wood, 1986.
3. **Back to Basics**, Reader's Digest, 1997.
4. **Joy of Cooking**, Rombauer and Becker, 915 pages, 1975 Edition or before. (The 1975 edition was reprinted for many years until at least 1986. When the book was finally changed from 915 pages, the really good information was replaced with quick recipes and microwave cooking.)
5. **New Illustrated Guide to Gardening**, Reader's Digest, 2000.
6. **Seed Sowing and Saving**, Turner, 1998.
7. **Big Book of Preserving the Harvest**, Costenbader, 2002.
8. **The Medical Advisor**, Time Life Books, Hardcover, 1996.
9. **Reader's Digest Guide to Medical Cures & Treatments,** Hardcover, 1996.
10. **Merck Manual of Medical Information**, 2nd Home Edition, 2004, Paperback, more than 1900 Pages, Highly Recommended.
11. **American Red Cross First Aid & Safety Handbook**, Kathleen A. Handal, M.D., 1992.
12. **DK First Aid Manual**, American College of Emergency Physicians, 2004.
13. **DK Natural Health Encyclopedia of Herbal Medicine,** Andrew Chevallier, 2000.
14. **The Complete Medicinal Herbal,** Penelope Ody, 1993.
15. **Reader's Digest Complete Guide to Sewing**, 1976 or 2002 Edition.
16. **Fundamentals of Building Construction**, 4th Ed., Allen & Iano, 2004.
17. **Complete Modern Blacksmith**, Weygers, 1997.
18. **New Complete Hoyle Revised,** Hardbound, 1991.
19. **The Complete Book of Solitare and Patience Games**, 225 Games, Morehead, 1983.
20. **Nelson's NKJV New King James Version Study Bible,** 2005 (or the **NIV Study Bible**).

I also suggest that you consider purchasing three of my other books:

1. **Grandpappy's Survival Manual for Hard Times, Third Edition.**
2. **How to Tan Animal Hides and How to Make High Quality Buckskin Clothing.**
3. **Grandpappy's Campfire Survival Cookbook.**

Index

About the Author

Robert Wayne Atkins, P.E. (Grandpappy)

Picture Taken in 2004.
Grandpappy Age 55.

Born in 1949. Accepted Jesus Christ as Savior in April of 1976.
B.S. Degree in Industrial Engineering & Operations Research,
Virginia Polytechnic Institute and State University,
Blacksburg, Virginia, June 1972.
Master of Business Administration, Major in Marketing,
Georgia State University, Atlanta, Georgia, March 1985.
Licensed Professional Engineer (P.E.), Florida 1980, Georgia 1982.
Ordained Deacon in Christian Church, Ocala, Florida, 1980.
Member of The Gideons International continuously since 1979.
Author of Nine Computer Software Games, including "**The Lost Crown of Queen Anne**," 1988-1991.
Contributing Author to "**Maynard's Industrial Engineering Handbook**," 5th Edition, p. 5.10, 2001.
Contributing Author to "**Maynard's Industrial & Systems Engineering Handbook**" 6th Edition, p. 102, 2023.
Listed in "**Who's Who in America**," 64th Edition, 2010.
Listed in "**Who's Who in the World**," 29th Edition, 2012.
Recipient of "**Who's Who**" Lifetime Achievement Award, 2019.

Robert is a descendant of the early European settlers in Virginia who married American Indian Cherokee wives.

In the year 2003 Robert began writing down what he had learned from the Holy Bible after more than 25 years of reading it on a daily basis. But he suspected his children might not read what he wrote if he put it into a sermon format. Therefore he began writing short Christian poems because he hoped that his children, grandchildren, and his other future descendants might take the time to read a short poem even if it was Christian oriented.

Later it occurred to him that some of his knowledge might be of interest to other people. Therefore he began publishing some of his writings on his website. Based on the favorable email feedback he received he continued to freely share his knowledge on his website. After several years, and at the request of many of his readers, he eventually consolidated some of his writings into the books below.

Books Written by Robert Wayne Atkins, P.E.

1. Handbook of Industrial, Systems, and Quality Engineering (English and Spanish).
2. Introduction to Engineering Management.
3. Engineering Economy and Financial Analysis (Student Textbook + Instructor's Manual).
4. Engineering Statistics and Applications (An Instructor's Manual is available).
5. Introduction to Quality Engineering (An Instructor's Manual is available).
6. Introduction to Industrial and Systems Engineering (An Instructor's Manual is available).
7. Work Measurement and Ergonomics (An Instructor's Manual is available).
8. Facilities Design and Plant Layout (An Instructor's Manual is available).
9. Practical Small-Scale Electrical Energy Systems.
10. Practical Strategies for Long-Term Survival.
11. The Practical Prepper's Survival Handbook.
12. Self-Defense Weapons: Traditional and Modern.
13. How to Maximize Your Eating Pleasure and Your Life Expectancy.
14. The Common Sense Diet.
15. The Food Book.
16. Grandpappy's Gourmet Cookbook.
17. Grandpappy's Recipes for Hard Times.
18. Grandpappy's Campfire Survival Cookbook (English and Spanish).
19. Grandpappy's Survival Manual for Hard Times, Third Edition.
20. The Most Important Survival Skills of the 1800s.
21. How to Tan Animal Hides and How to Make High Quality Buckskin Clothing.
22. How to Live Comfortably for Several Years in a Hostile Wilderness Environment.
23. The New Heaven and The New Earth (English and Spanish).
24. Some Difficult Questions Answered Using the Holy Bible.
25. Religion and Christianity in the Twenty-First Century.
26. Grandpappy's Christian Poems.
27. Grandpappy's Stories for Children of All Ages.
28. Ancient Board Games and Solitaire Games from Around the World.
29. The Four Pillars of Prosperity: Governments, Businesses, Religions, and Banks.

Made in United States
Cleveland, OH
05 November 2026

47649165R00045